EVERYDAY SKETCHING and DRAFTING

SECOND EDITION

J. W. GIACHINO

*Head and Professor Emeritus
of Engineering and Technology
Western Michigan University
Kalamazoo, Michigan*

H. J. BEUKEMA

*Professor of Engineering Graphics
Department of Engineering and Technology
Western Michigan University
Kalamazoo, Michigan*

 American Technical Society • **Chicago 60637**

Preface to the Second Edition

Many opportunities for today's labor force can be found in the fields of engineering, applied science, and technology. It is within these fields that sketching and drafting find their widest application. In the building trades, the manufacturing, processing, and electronics industries, millions of workers must be able to interpret and work from building plans, blueprints, and electrical diagrams.

The ability to interpret sketches and drawings, however, is not limited merely to those directly connected with the end products of construction or production. Added millions of people who are indirectly or even remotely connected with the various fields find the knowledge of drafting important to their business interests. Inventors, designers, investors, trade editors, company managers, sales and service personnel must all have a working knowledge of blueprints, drawings, and plans in order to correctly handle various responsibilities to their clients and customers. Often, a stepping-stone to positions of greater responsibility, in and out of industry, is the ability to interpret and often to make various sketches and drawings.

Drafting, like mathematics, is a medium of precise communication. Anyone who cannot afford the chance of being mis-understood is likely to use a sketch or drawing to help convey the idea. Professional people, sales people, management personnel, technicians, and average people alike, at one time or another use the language of drawings to more forcefully communicate ideas.

This book is an introduction to the field of drafting. It is designed to fit a wide range of applications. The student will find that much of the content applies to everyday situations and is therefore useful, even if he does not intend to pursue the subject of drafting beyond this initial course. To the student intending a career in drafting, the trades, or some related field, this course will provide a solid grounding in the fundamentals of drafting.

This Second Edition includes over 75 pages of self-check questions and drawing problems. Teachers may thus assign as many problems as time permits. Self-check questions and drawing problems have been carefully selected to reinforce the important material of each unit. Problems ranging from the simple to the complex allow the teacher to select problems which will be challenging and instructive to students whatever their level of achievement. A new answer key to self-check questions is provided at the end of the book so students can check on their individual progress.

Units which include material on architectural drawing, sheet metal drawings, machine-shop drawings, and electrical diagrams assure that the material will have carry-over value to other courses.

The authors, J. W. Giachino and Henry J. Beukema, have worked very closely with industry and have been eminently successful in identifying trends in industry. Having previously collaborated on several text books for the field, this highly successful team is thoroughly familiar with the needs of, and the problems faced by, the student in his initial approach to the field of drafting. This book reflects the wide experience of the authors in their connections with industry and in their capacities as educators.

The Publishers

Contents

PAGE

UNIT 1. Communicating with Sketches and Drawings 1

What are sketches, diagrams, and mechanical drawings? Who uses sketches, diagrams, and mechanical drawings?—Scope of this course.

UNIT 2. Learning to Sketch 7

Sketching paper—Pencils—Sketching horizontal lines—Sketching vertical and slanted lines—Sketching squares and rectangles—Sketching angles—Sketching circles and arcs—Sketching irregular curves—Sketching a hexagon—Sketching an ellipse—Sketching a one-view drawing.

UNIT 3. Sketching Views 21

Meaning of views—Number of views—Arrangement of views—Projection—Alphabet of drawing lines—Rounds and fillets—Sectional views.

UNIT 4. Lettering 39

Forming letters and numbers—Spacing letters and words—Height of letters and words—Height of numerals.

UNIT 5. Dimensioning 47

Dimensioning a drawing—Placement of dimensions—Dimensioning circles, arcs, and angles—Notes—Decimal dimensioning—Tolerances.

UNIT 6. Showing Fasteners on Working Drawings 61

Working drawings — Threaded fasteners — How threads are represented — Types of threaded fasteners—Types of non-threaded fasteners.

UNIT 7. Sketching Pictorial Drawings 77

Isometric drawings—Oblique drawings—Perspective drawings.

UNIT 8. Making Instrument Drawings 93

Drawing board — Pencils — T-square — Triangles — Protractor — Compass — Dividers —Curves—Measuring scales—Decimal scale—Drawing an object to scale.

UNIT 9. Drawing Geometric Constructions 115

Bisecting a line—Bisecting an arc—Bisecting an angle—Dividing a line into equal parts —Drawing an arc tangent to two lines at 90°—Drawing an arc tangent to two lines not at 90°—Drawing an arc tangent to a straight line and an arc—Drawing tangent arcs—Drawing a straight line tangent to two arcs.

UNIT 10. Architectural Drawing 121

Types of house architecture—Sketching preliminary floor plans—Finished set of architectural plans—Building specifications.

UNIT 11. Drawing Charts 143

Line charts—Bar charts—Circle charts.

UNIT 12. Project Drawing and Print Reading 155

Planning the project—Duplicating a design—Sheet metal drawings—Electrical diagrams.

Answer Key 172

Communicating with Sketches and Drawings

From the time man came on earth, the ability to communicate has been considered essential in all walks of life. Civilization, as we know it today, could never have come into existence without some form of communication.

Most of us are inclined to consider communication as a means of conveying to our associates our ideas, wants, or beliefs, either by written or spoken words. And to a large extent this is true. As a matter of fact, from the first day you started school you have been taught to read, write, and speak so that you can communicate with others.

The art of communication, however, has further meanings. We know that on some occasions our thoughts or messages can be more effectively conveyed if we use other media of expression. For example, remember the day you tried to explain to friends the way to your home. They could not quite understand your directions, so you took a piece of paper and sketched a rough map, showing them how to proceed. See Fig. 1. Instead of telling them with words, you used a diagram. Actually, for modern living, we must often rely on various graphical representations, in addition to the written and spoken word, to communicate effectively with our associates.

What are sketches, diagrams, and mechanical drawings?

Quite often the terms *sketches*, *diagrams* and *mechanical drawings* are used synonymously to designate any form of graphical representation. In some respects they are closely related. Specifically they could be defined in this way:

Sketches. A drawing of an object made freehand is called a sketch. The sketch may show the object as it is normally seen in a photograph, as in Fig. 2, or by several views placed in certain positions, as in Fig. 3.

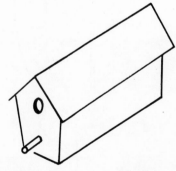

Fig. 2. A freehand sketch may represent an object in pictorial form.

Diagrams. A drawing which uses lines and symbols to convey directions, to illustrate related sets of information on a graph or chart, or possibly to show the wiring of an electrical circuit, is called a diagram. See Fig. 4.

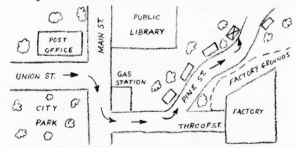

Fig. 1. A diagram helps to convey your directions more clearly.

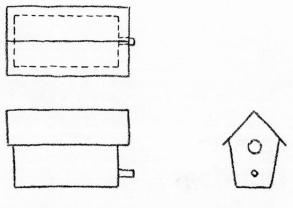

Fig. 3. This freehand sketch shows the outline of the object in Figure 2 when viewed from different positions.

Mechanical drawings. An exact representation of an object which is prepared with drawing instruments is called a mechanical drawing. Such drawings are used in the construction and manufacture of any product, whether it is a simple item such as a safety pin or a complex structure such as a missile. See Fig. 5. The people who prepare these drawings are called draftsmen. See Fig. 6.

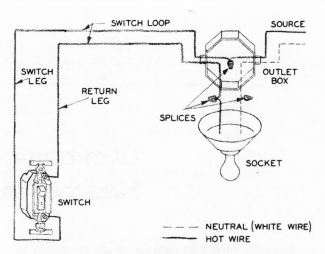

Fig. 4. This example of a diagram shows a ceiling light and a switch hook-up.

Who uses sketches, diagrams, and mechanical drawings?

Have you ever stopped to consider how often people must rely on some kind of a drawing to accomplish what they want to do? Or have you noticed how frequently books, magazines, newspa-

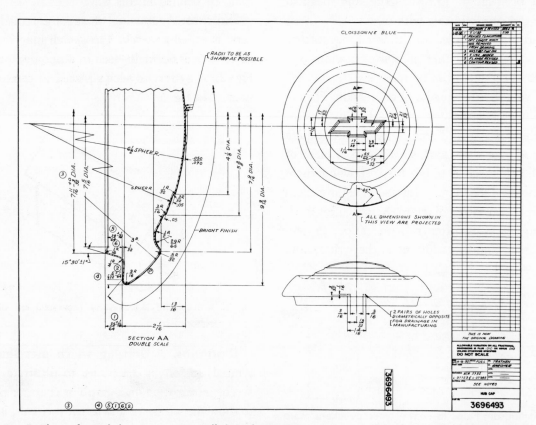

Fig. 5. This industrial drawing contains all the information necessary to manufacture the hub-cap shown.

Fig. 6. The man in industry who prepares mechanical drawings is known as a draftsman.

pers, and even television programs utilize drawings, sketches, and diagrams to get a message across to their readers or listeners?

Drawings are often found to be vital to all types of sports, professions, and occupations. In sports, football and basketball coaches rely extensively on diagrams when demonstrating a new play to their teams. The players use the same diagrams to memorize the plays. With diagrams, the players know exactly what is expected of them. Fig. 7 shows a basketball play used against a man-to-man defense.

In the professional field, doctors and medical technicians must be able to interpret drawings and diagrams which tell how to use the complex mechanical and electronic devices of today's medicine, such as iron lungs, physical therapy machines,

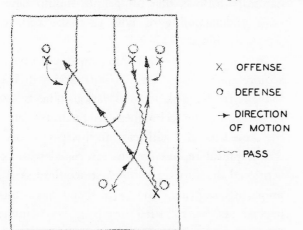

X OFFENSE

○ DEFENSE

→ DIRECTION OF MOTION

〰 PASS

Fig. 7. A coach often sketches a diagram to illustrate a play in basketball.

artificial limbs and organs, x-ray and electrosurgical equipment. Lawyers must be able to explain charts and diagrams to a jury and to their clients. Bankers and accountants must have the ability to obtain necessary facts and figures from graphs.

In all industries, drawings are the essential means of communication. Without drawings our modern industries could not exist. Were it not for the drawings turned out by the draftsmen, our mass production methods would be impossible. Machinists, production line foremen, carpenters, plumbers, indeed all workers in the building and production trades, must be able to read and interpret drawings. The jobs of today demand precision and skill on the part of the worker to adequately meet the needs of mass production and assembly methods.

If you stop to realize that behind every piece of machinery there is at least one drawing, you begin to see the importance of this medium of communication. Consider for a moment our satellites, missiles, battleships, diesel locomotives, automobiles, and electrical appliances. Each of these is composed of many, many parts, and for every part there is at least one drawing. See Fig. 8.

Drawings are recognized the world over as the graphical language to represent ideas, design, or construction. They are means of expression used by the scientist, engineer, designer, technician and tradesman. Regardless of their work these people either make sketches and drawings or must be able to read them. Usually an idea starts with a rough sketch, then the sketch is refined and eventually it is made into a finished mechanical drawing.

There are times, then, when everyone must make some kind of a sketch or read some kind of a drawing. Without knowing something about drawings you would find it almost impossible to do correctly the many things which you must do in all of your daily activities.

Whether a man installs a fixture, adds improvements to, or makes repairs on, his house he usually requires a sketch or drawing to help him. Women also must resort to helpful drawings, whether the chore be cutting out a pattern for a dress or setting

Lincoln-Mercury Division—Ford Motor Co.

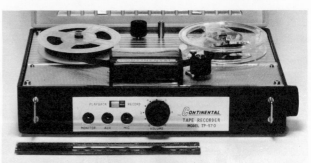

Continental Sound Engineering Co.

American Radiator and Standard Sanitary Corporation

Boeing Co.

United States Air Force

Fig. 8. The manufacture of these products required hundreds of drawings.

up baby's folding play pen. Without the simple sketch or drawing, most Do-It-Yourself projects would be difficult indeed.

Radio kits, toys, electric hedge clippers, light fixtures, door chimes, power lawnmowers, outboard engines, hobby kits, and thousands of other items come with assembly, installation, and/or maintenance and repair instructions. In each case there is usually a drawing or sketch that shows you how the various parts of the object are to be assembled, installed, cleaned, lubricated, cared for or maintained.

Scope of this course

What we have tried to convey so far is that drawings of one form or another play an impor-

tant role in the lives of all people. If this is so, it naturally follows that all people should have a basic understanding of how to make and read simple drawings.

In this course you will have an opportunity to acquire such knowledge. To start, you will learn to make freehand sketches of various objects based on principles and techniques used by draftsmen. At the same time you will learn to interpret many of the graphical representations associated with conventional drawings. Having first acquired skill in preparing sketches, you will then make more precise drawings with drafting instruments.

Check your knowledge of this unit by completing the Self-Check for Unit 1 on pages 5 and 6.

SELF-CHECK FOR UNIT 1

PART 1

DIRECTIONS: In the space provided answer each question as briefly as possible. Use complete sentences. *Self-Check answers may be found on page 172.*

1. Why are drawings considered a medium of communication?

2. What is the difference between a sketch and a mechanical drawing?

3. Why are drawings important in industry?

4. What is the simplest type of drawing?

5. What is the most complex type of drawing?

6. Why are different types of drawings necessary?

6

7. Why do most people need a basic knowledge of making and reading drawings?

8. List ten objects that are usually assembled at home by the buyer.

PART 2

Identify the type of drawing for each of the illustrations which appear below. Write each name beneath the illustration. *Self-Check answers may be found on page 172.*

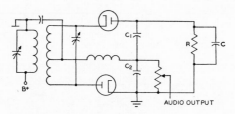

1. _____

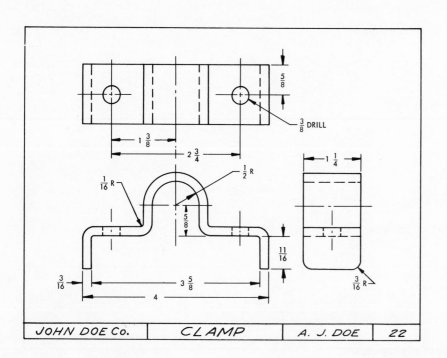

JOHN DOE Co. | CLAMP | A. J. DOE | 22

2. _____

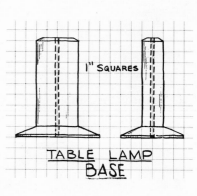

1" SQUARES

TABLE LAMP BASE

3. _____

Score_____

Learning to Sketch

The ability to sketch is something that you can acquire through practice and by following certain basic sketching techniques. In this unit you will have an opportunity to study and apply these techniques.

Sketching paper

At first, it is usually better to use cross-section paper for sketching. As you gain skill you will be able to make reasonably good sketches on plain paper.

The grid lines on cross-section paper will help secure good proportions and serve as guides in

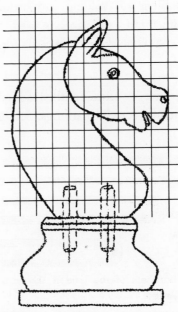

Fig. 2. Sketching is simplified when cross-section paper is used.

sketching lines. The squares can be used to represent certain sizes. By counting either horizontally or vertically, the correct shape of the object can readily be sketched. Thus each square can represent ⅛″, ¼″, ½″, 1″, or any other convenient unit. See Figs. 1 and 2.

Cross-section or grid paper, as it is sometimes called, usually is 8½ by 11 inches in size. Selection of square size depends on how large or small you want to make your sketch. Naturally the larger the sketch, the greater should be the square size. Incidentally, you will have to guess distances that fall in between the designated square size. Thus if a dimension is 1¾ inches and each square

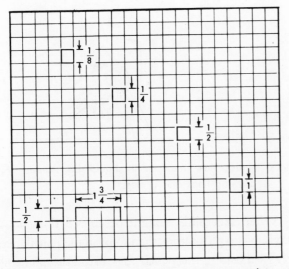

Fig. 1. Squares can be used to represent a variety of sizes.

is considered to be ½ inch, you will need to use three squares and one-half of another one. See Fig. 1.

Maintaining correct proportions is important in freehand sketching. Regardless of how excellent the sketching techniques may be, the final sketch will not be very good if it has poor proportions. Proper proportioning is obtained by estimating actual dimensions. Sketches are not made to scale and you have to learn to recognize proportions by comparison. Thus if the height is twice the width, then this same proportion can be maintained for other details.

The problem of holding proportions is not quite as difficult if grid paper is used. More care must be taken when sketching on plain paper.

Pencils

The hardness or softness of the lead is designated on the pencil by certain markings. See Fig. 3. The

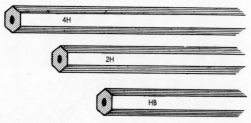

Fig. 3. The hardness of the lead is shown on a pencil by symbols.

grades run from 7B which is very soft, through 6B, 5B, 4B, 3B, 2B, B, HB, F, H, 2H, 3H, 4H, 5H, 6H, 7H, 8H, and 9H, which is the hardest. The best pencil for freehand sketching is one with a medium (F) or soft (HB) lead.

It is important to obtain the proper point on the pencil. Pencils are usually sharpened in a pencil sharpener. However, a pencil may be sharpened with a knife if the point is then filed to conform to the dimensions shown in Fig. 4.

Hold the pencil loosely approximately 1½ to 2 inches from the point as in Fig. 5. The practice is to slant the pencil at an angle of 50 to 60 degrees from the vertical for drawing straight lines and at about 30 degrees for circles. See Fig. 6.

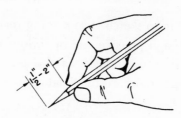

Correct (A)   Incorrect (B)

Incorrect (C)

Fig. 4. A pencil must be sharpened correctly to produce good sketches.

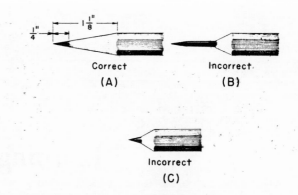

Fig. 5. Hold the pencil as shown for sketching.

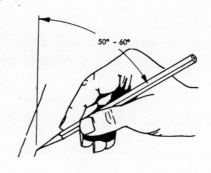

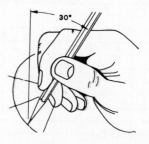

Fig. 6. Notice how the pencil should be slanted for drawing lines and circles.

When sketching straight or curved lines, it is advisable to pull rather than to push the pencil. See Fig. 7. Pushing the pencil may cause the point to catch the surface of the paper and puncture it. As

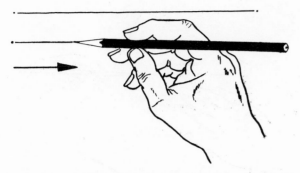

Fig. 7. When sketching straight or curved lines always pull the pencil.

the pencil is pulled, it is a good idea to rotate it slightly, since this motion keeps the point sharper for a longer period. When the line begins to widen, the pencil needs resharpening.

Sketching horizontal lines

To draw horizontal lines, first mark off the end points with dots to indicate the position of the line. Then sketch the line between the two points, moving the pencil from left to right. For long lines use a series of dots for better alignment. See Fig. 8.

Short lines are best drawn with a finger and wrist movement. As the line becomes longer, it is better to use a free arm movement, since the fingers and wrist tend to bend the line.

Before actually drawing the line, it is often a good idea to make a few trial swings between the points with the pencil held about one quarter-inch

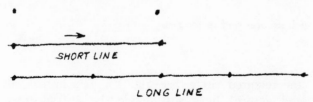

SHORT LINE

LONG LINE

Fig. 8. Use this procedure in sketching short and long lines.

above the paper. This helps you to adjust the eye and hand for the intended line. Next sketch a very light line between the points with an eye on the point where the line is directed. Then darken the trial line. With practice the initial light trial line may become unnecessary.

Sketching vertical and slanted lines

Sketch vertical lines by starting at the top end of the line and moving the pencil downward. Slanted lines can be sketched better if the pencil is moved from left to right. See Fig. 9. The beginner may wish to treat all lines as horizontal lines by turning the paper as shown in Fig. 10.

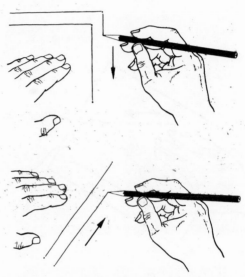

Fig. 9. When sketching vertical lines, move the pencil in a downward direction; move to the right when sketching slanted lines.

Sketching squares and rectangles

To sketch a square, draw horizontal and vertical center lines as shown in A of Fig. 11. Space off equal distances on these lines as in B of Fig. 11. Sketch light horizontal and vertical lines through the outermost points to form the square. Then darken the lines as in C. Use a similar procedure to sketch rectangular shaped objects. See D in Fig. 11.

Sketching angles

To sketch an angle other than 90°, first draw two lines to form a right angle. Angles are formed by intersecting lines or lines that meet at a point. To sketch an angle smaller than a right angle

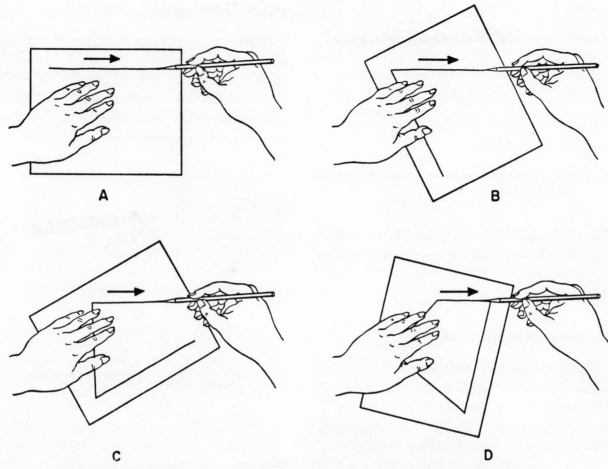

Fig. 10. It is often easier to turn the paper so all lines can be sketched in a horizontal position.

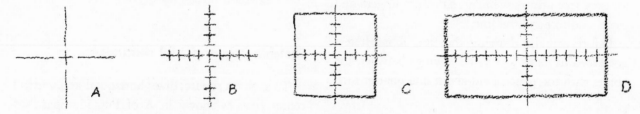

Fig. 11. Use centerlines to sketch squares and rectangles.

(90°), lightly draw equally spaced lines which will divide the right angle into several equal and smaller angles. See Fig. 12. The most commonly used angles are the 15, 30, 45 and 60 degree angles. Project a dark line through the one point that represents the angle desired.

Sketching circles and arcs

To sketch a circle, draw a horizontal and a vertical line through a point making the center of the required circle. See Fig. 13. The horizontal and vertical lines will divide the 360 degrees about the point into four quarters, or quadrants, each containing 90 degrees. On a piece of scrap paper, mark off the desired radius. Transfer this distance to the centerlines by holding one end to the point which represents the center of the circle and marking off the desired length along the centerline. The circumference of the circle is drawn by finding enough points to accurately determine

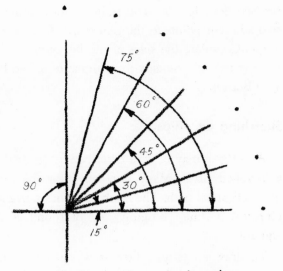

the curve of the circle. Therefore, draw diagonal lines at various intervals through the center and locate these other radial points, using the length marked on the scrap paper.

Complete the 360° circle by drawing short arcs through one quadrant at a time. By rotating the paper, the stroke can be made in the same direction each time, until the entire circumference is drawn.

Sketching irregular curves

For an irregular curve, locate a number of points

Fig. 12. Correct way to sketch angles.

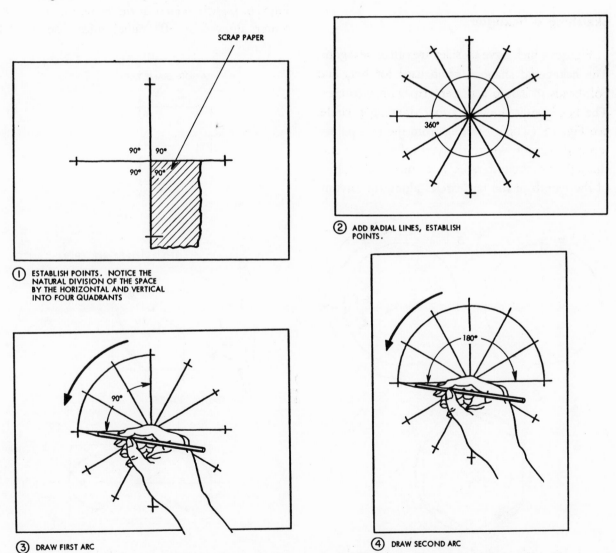

Fig. 13. Make your strokes in the same direction when drawing the circumference.

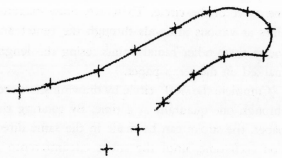

Fig. 14. The irregular curve is drawn as a series of curves.

to represent the shape of the required curvature. Complete the curve by drawing a series of arcs through these points as in Fig. 14.

Sketching a hexagon

Figures which have six sides are called hexagons. The hexagonal shape is often used for nuts and bolt-heads to assure a firmer grip with wrenches. The hexagon is drawn by first drawing a circle. See Fig. 15 (Top). Starting from the two points, top and bottom, where the vertical axis line cuts through the circumference line, mark off a third of the length of the half-circles along the circum-

ference. See Fig. 15 (Top right). Lines from any two adjacent points to the center will form a 60° angle. Complete the sides of the hexagon by connecting the six points with straight lines. See Fig. 15 (Bottom).

Sketching an ellipse

An ellipse may best be described as an oval or a flattened circle. Many satellites orbit the earth in an elliptical path. The earth itself has a slightly elliptical outline, "bulging" out of round at the equator.

To draw an ellipse, first draw the major and minor diameters as centerlines of a rectangle. See Fig. 16. Lightly sketch in the rectangle. From the center, draw four 30° radial lines, one in each

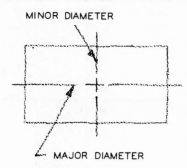

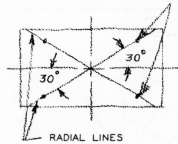

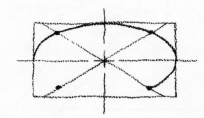

Fig. 16. Points on radial lines will determine the curve of the ellipse.

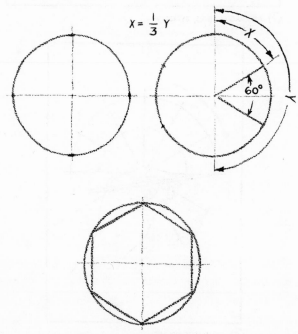

Fig. 15. The circle is divided into six equal angles in sketching a hexagon.

quadrant. On the four radial lines mark off points equally distant from the center; the ellipse will pass through these points. Sketch in the outline of the ellipse. Different points may be plotted on additional sets of radial lines when greater accuracy is needed.

The ellipse, like the circle, is symmetrical about both axes. An object is symmetrical with respect to an axis if *corresponding* points on opposite sides of the axis line are equally distant from the axis. The axis lies between the corresponding points. An object may be symmetrical to one axis or it may be symmetrical to both axes. See Fig. 17.

Notice in Fig. 17 (Top) that the ellipse is symmetrical with respect to its vertical axis, but the quadrangle is not symmetrical. In Fig. 17 (Bottom) the circle is symmetrical with respect to both the horizontal and vertical axes, while the triangle can be symmetrical with respect to its vertical axis only. A good rule of thumb to remember regarding symmetrical objects is that if one half of the object seems to be a reflection of the other half in an imaginary mirror placed along the axis line, it is symmetrical.

Sketching a one-view drawing

A one view drawing is simply a flat lay-out or outline of an object as seen when looking squarely at one surface. The objects shown in Fig. 18 are typical examples of a one-view drawing.

To sketch a flat layout, draw base lines, or centerlines, and locate centers of arcs and circles.

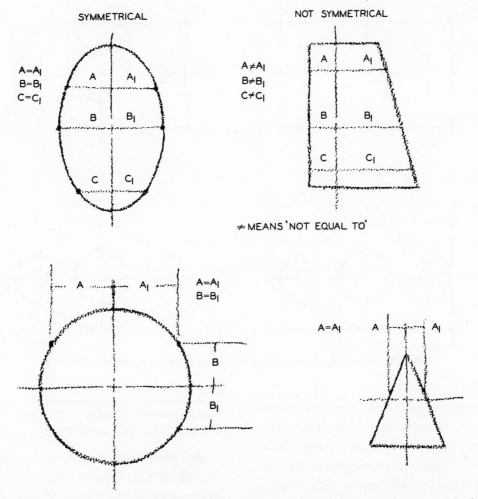

Fig. 17. Objects may be symmetrical with respect to the horizontal axis, the vertical axis, or to both axes.

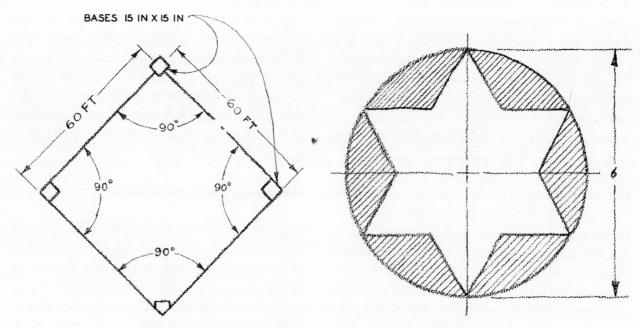

Fig. 18. A one-view drawing shows only a single view of an object.

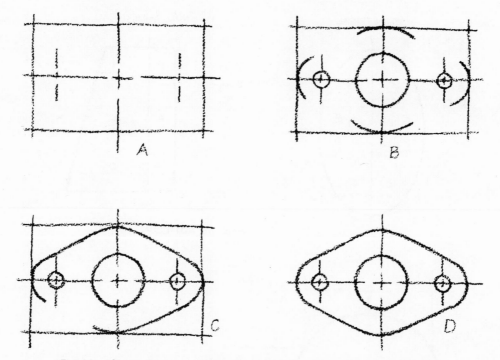

Fig. 19. Always draw curved lines first when sketching a one-view drawing.

Block in the outline with light lines. Always draw circles and arcs first and then draw straight lines. You will find it easier to secure smooth connections if straight lines are drawn to the arcs than by trying to make arcs meet the straight lines. See Fig. 19.

Check your knowledge of this unit by completing the Self-Check for Unit 2 on pages 15 and 16.

Practice making sketches of the objects on Drawing Sheets 1, 2, 3 and 4, pages 17-20. Follow the instructions on the bottom of each sheet.

SELF-CHECK FOR UNIT 2

PART 1

DIRECTIONS: Complete the following statements by writing the correct word or words in the space provided. *Self-Check answers may be found on page 172.*

1. The best pencil for sketching is one with a hardness symbol of _____ or _____.

2. For sketching straight lines, slant the pencil _____ to _____ degrees from the vertical.

3. Circles are drawn more easily if the pencil is slanted about _____ degrees.

4. In sketching, the pencil is held loosely and about _____ inches from the point.

5. Horizontal lines should be drawn by moving the pencil from _____ to _____.

6. Vertical and slanted lines should be drawn by starting at the _____ of the line and moving the pencil _____.

7. The first step in drawing either a square or rectangle is to draw a _____ line and a _____ line that cross near the middle.

8. A right angle has _____ degrees.

9. Each quarter of a circle contains _____ degrees.

10. A hexagon has _____ sides.

11. A one-view drawing merely shows a _____ layout of an object.

12. In sketching a one-view drawing, always draw _____ first and then draw _____ lines to them.

PART 2

DIRECTIONS: The following statements are either True or False. Draw a circle around the T if the statement is true or around the F if the statement is false.
Self-Check answers may be found on page 172.

1. T F The hardness or softness of lead ranges in grades from 7B, which is the softest, to 9H, which is the hardest.

2. T F When sketching lines, it is always best to push rather than pull the pencil.

3. T F When sketching horizontal lines the pencil should always be moved from left to right.

4. T F When drawing square or rectangular objects, it is advisable to first draw centerlines.

16

5. T F The most commonly used angles are the 15, 35, 55 and 70 degree angles.

6. T F When drawing a circle, diagonal lines are drawn through the center point and marked with the length of the radius.

7. T F When drawing the circumference of a circle, try not to rotate the paper.

8. T F A hexagon has eight equal sides.

9. T F Circles and ellipses are symmetrical about the horizontal and vertical axes.

10. T F An object can be made symmetrical to both axes.

11. T F When sketching one-view drawings, omit centerlines and draw all straight lines first.

12. T F When making a sketch on cross-section paper, the size of the squares used regulates the size of the sketch desired.

Score_____

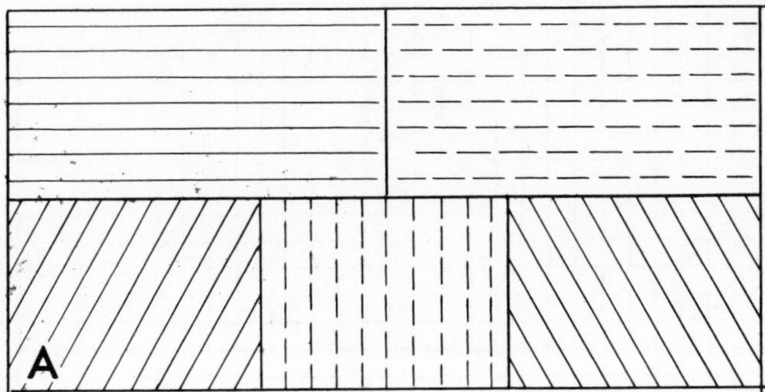

A

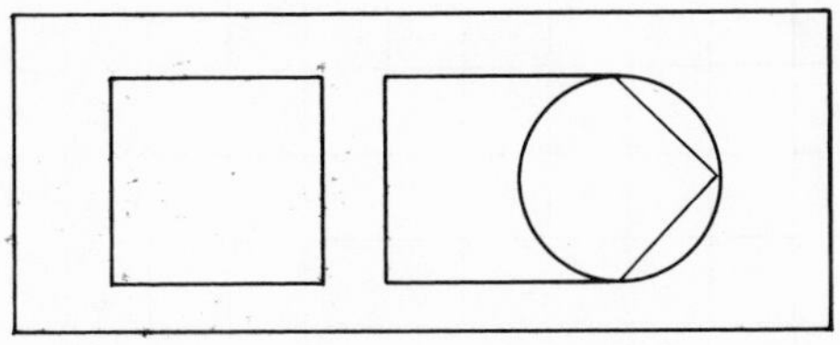

B

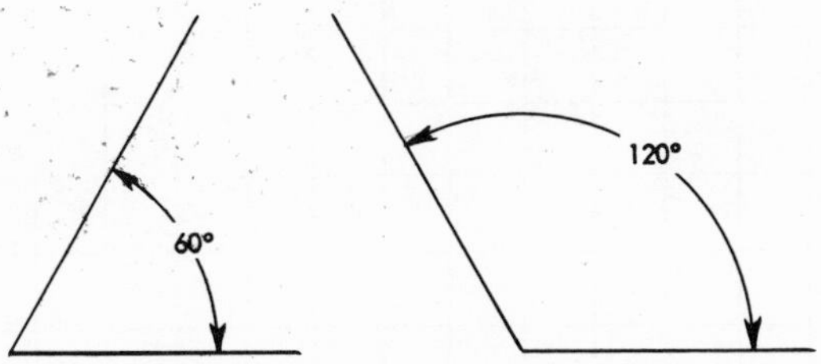

C

60°

120°

D

SCHOOL

E

Make freehand sketches of A, B, C, D, and E in the spaces provided.

TITLE		DRAWING NO.
NAME	GRADE	

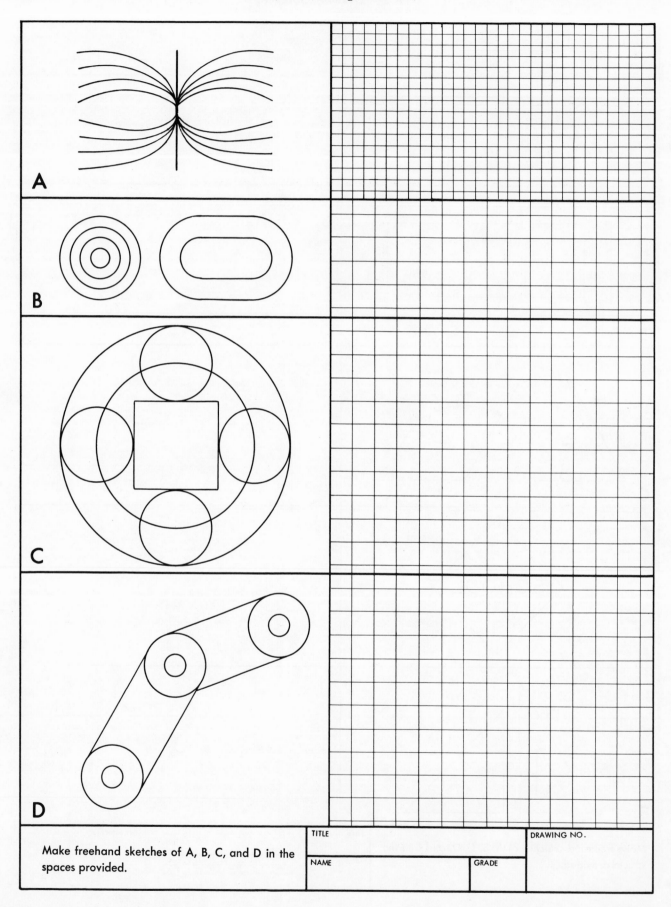

A

B

C

D

Make freehand sketches of A, B, C, and D in the spaces provided.

TITLE

NAME

GRADE

DRAWING NO.

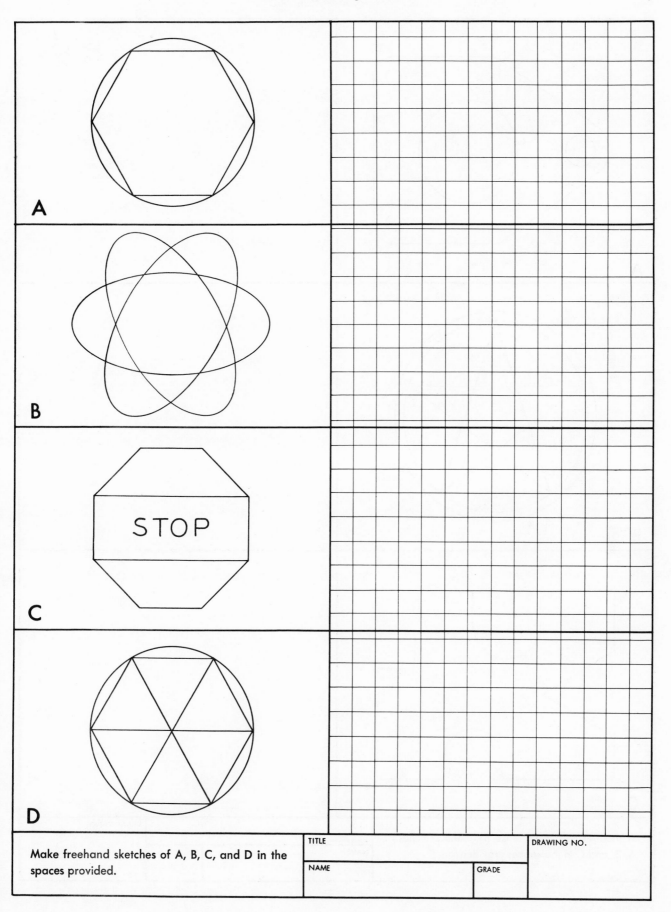

Make freehand sketches of A, B, C, and D in the spaces provided.

TITLE		DRAWING NO.
NAME	GRADE	

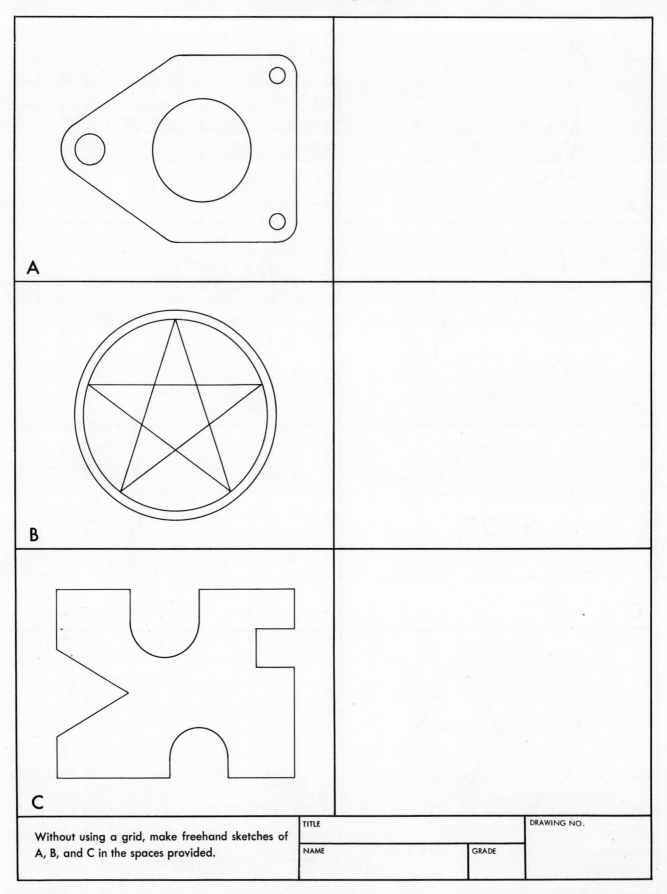

A

B

C

Without using a grid, make freehand sketches of
A, B, and C in the spaces provided.

TITLE		DRAWING NO.
NAME	GRADE	

Sketching Views

In the previous unit you learned to lay out an outline or what is better known as a one-view drawing of an object. Very often a single view is not enough to accurately describe the true shape of the object. Therefore, additional views must be included so that the object can be readily constructed. A drawing with several views is known as a multiview drawing. This system of views is the basis for all industrial drawings.

Meaning of views

To get a better understanding of what is meant by views, examine the block shown in Fig. 1. If you look at this block directly in front, you will find that it appears as a rectangle. The shape as seen from such a position is known as *the front view*.

Now if we want a clearer picture, we also take a look at the block from above. Here too the block appears as a rectangle. The shape as seen from here is called *the top view*.

Usually we need still another view to get a complete picture. This time we look at the object from the side; the block appears L-shaped. The shape as we see it from here is known as *the side view*. In most cases the right side view is used. There are some occasions when the left rather than the right side of the object is drawn. This is done only when the left side presents a clearer picture.

Another way to visualize views is to imagine that you are holding the object in your hand as shown in Fig. 2. By looking directly at the front of the object you can readily see what the front view should look like. Similarly, by looking at the top and side of the object you can see how these views appear.

As a rule, the three views mentioned are those most frequently used for a multiview drawing. Sometimes the shape of the object requires that the bottom, rear, and both sides be shown. These additional views are drawn only when the shape of the piece is very complicated and the added views are essential.

Fig. 1. The principal views of an object for a multiview drawing are the three shown.

Number of views

A drawing should include only the views that

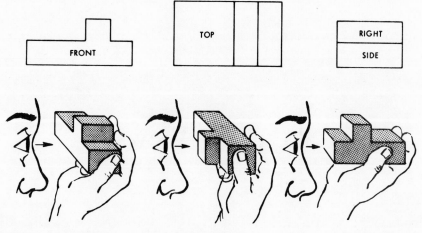

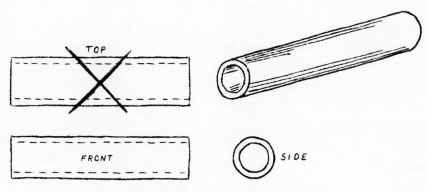

Fig. 2. Notice how each view appears when looking directly at the three principal surfaces of this block.

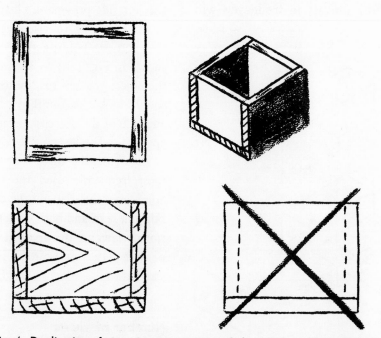

Fig. 3. Only two views are needed for this tube.

Fig. 4. Duplication of views is unnecessary and should always be avoided.

are absolutely necessary. In other words, if one view is an exact duplication of another, then such a view is omitted. Take for example the tube shown in Fig. 3. Here only a front and side view are required. A top view would merely be a duplication of the front view. The same is true of the box in Fig. 4. Notice that only a front and top view are essential to show its true shape.

Arrangement of views

If views are to convey the exact shape of an object, they must be shown in certain positions. These locations are governed by the planes in which the views fall when the sides of the object are unfolded and spread in a flat position.

To understand better what is meant by placement of views, assume that the object is enclosed in a transparent plastic box and that its sides are hinged so they can be swung open. See Fig. 5. Now if each side of the plastic box is considered a plane to which a specific view of the object is projected, then when the sides are swung open each view falls into a definite position relative to the other views. Thus the top is directly above the front view, the sides are to the immediate right and left of the front and the bottom is below the front. However, as we indicated before, for most objects only the front, top, and right side views are used.

In making a multiview drawing, the views are spaced apart so that sizes can be inserted. You will learn more about placement of sizes later.

As you will see in Fig. 6, the three principal views are in their proper relationship to each other. If you follow the letters assigned to the

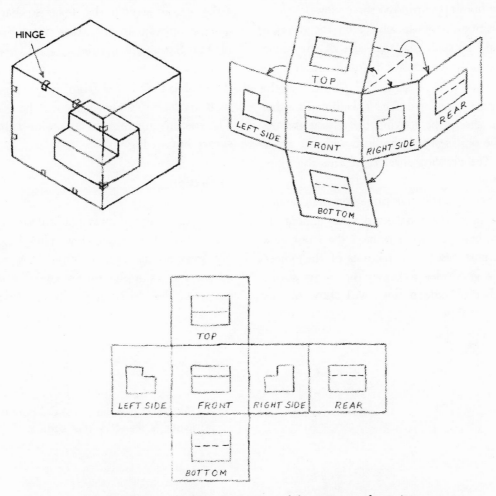

Fig. 5. All views unfold as shown around the stationary front view.

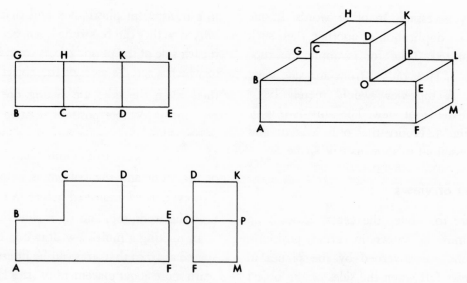

Fig. 6. Views spaced apart so sizes can be inserted.

pictorial block on the right, you can easily see where these same lines appear in the three views.

The positioning of the object on the drawing sheet should be such that the sides of the object having the most descriptive features are at right angles to your line of sight and parallel to the plane of projection. See Fig. 7 below. In other words, you should be looking directly at the object and the object should be flush with the drawing sheet. The drawing sheet represents the plane of projection.

The object should be positioned on the drawing sheet so it produces a balanced arrangement. In general the best practice is to have the front view show the most descriptive features of the object. Also it is a good idea to orient the object so the least number of hidden lines will show on the

views. Another point to keep in mind for good balance is to arrange the object so that its longest surface appears in the front view. Notice in A of Fig. 8 the unbalanced effects when the long view is shown in the side view. On the other hand if the object is viewed from the position indicated in B, too many hidden lines will have to be used. The best arrangement is to position the object as shown in C of Fig. 8.

Projection

In preparing a drawing, a draftsman first draws the outline of the front view. Then he projects the necessary points from the front view to the other views. For example, in preparing a drawing or sketch of the object in Fig. 9, the first step is to

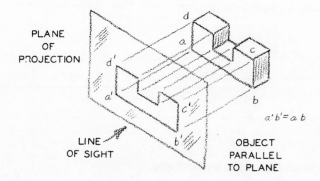

Fig. 7. The object and the plane are parallel; the plane image is therefore true size.

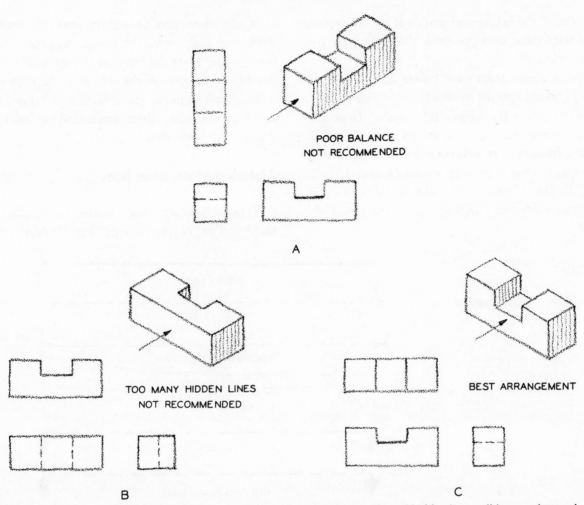

POOR BALANCE
NOT RECOMMENDED

A

TOO MANY HIDDEN LINES
NOT RECOMMENDED

BEST ARRANGEMENT

B   C

Fig. 8. Position the views of the object on the sheet so that the least number of hidden lines will have to be used.

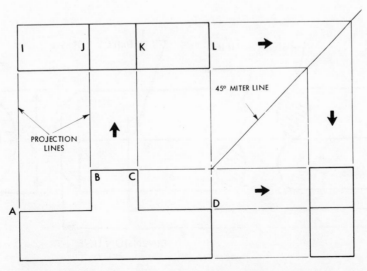

Fig. 9. Projection of views in this manner simplifies the drawing process.

determine the height and length of the front part, and with these sizes you draw the outline for the front view.

By using the front view, points A, B, C, and D are projected upward by means of projection lines. This transfers the object relationship from the front view to the top view. Points A, B, C, and D in the front view now become a part of the object lines in the top view, represented by lines I, J, K, and L. The width of the object in the top view is determined by measurements from the actual object.

In the same way projection lines are used to draw the side view. Extending projection lines horizontally from the front view will produce the height and features of the side view. Also by projecting lines from the top view to a 45° miter line and then extending them downward provides the width of the side view.

Alphabet of drawing lines

Various types of lines are used to describe the shape and size of an object. See Fig. 10. Notice that

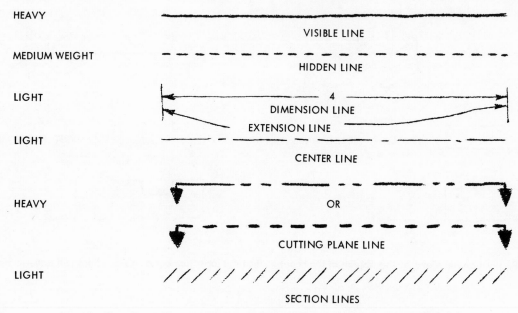

Fig. 10. The types of lines used in making a multiview drawing vary in thickness.

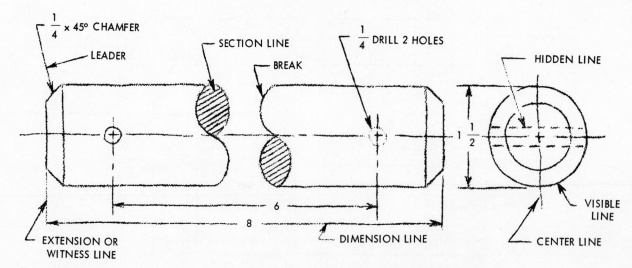

Fig. 11. How the various lines are used is shown in this typical drawing.

there are three weights of lines: heavy, medium, and light. Contrast in line weights is obtained not by degree of darkness but differences in thickness.

When making a drawing, keep all lines clean and sharp. Be sure they meet squarely at corners and blend smoothly at all points of contact. As soon as a line begins to vary in thickness, re-sharpen the lead.

Figure 11 illustrates how these lines are actually used in a drawing.

Visible lines. Heavy lines used to outline the exterior contour of the object are called visible lines. When objects have curved surfaces, these surfaces may appear as curves on some views and as straight lines on others. For some objects, curved lines will be seen in all views. See Fig. 12.

Hidden lines. Medium weight lines used to describe edges that are not visible to the eye are called hidden lines. It is slightly thinner than the visible line.

In drawing the various views of an object, there will be some surfaces which cannot be seen. Hid-

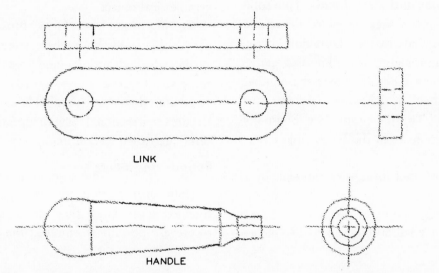

Fig. 12. Often curved surfaces will appear in other views as straight lines or circles.

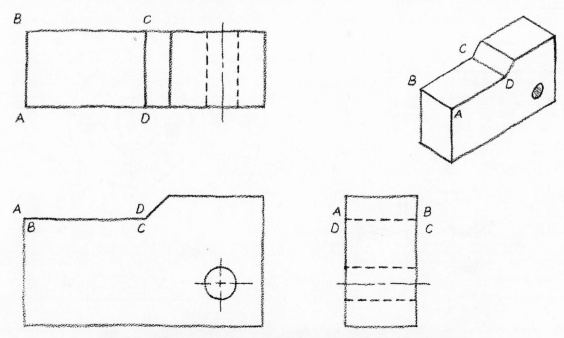

Fig. 13. Hidden features of an object are identified with hidden lines.

den lines are used to identify these surfaces. Notice in Fig. 13 that the hidden line in the side view represents edges AB and DC, or the surface ABCD. Also the hole is shown in both the top and side views with hidden lines.

The short dashes which form the hidden lines should be uniform in size. The practice is to make the dashes about 1/8″ to 3/16″ in length with a 1/32″ space between dashes. Usually a hidden line begins with the dash in contact with the object line.

Dimension lines and arrowheads. Thin solid lines used to designate sizes are called dimension lines. Numbers are inserted in the middle of the line to show length, width, or thickness. A gap is left in the otherwise solid line where the number is inserted. Numbers, or dimensions, are discussed in the next unit. The dimension line is usually placed 3/8″ or more from the heavy object line. See Fig. 14 (Top).

Arrowheads are used to indicate the ends of a dimension line. Arrowheads should be approximately three times as long as they are wide. For most drawings, they will average about 1/8 inches

in length. See Fig. 14 (Bottom) for the correct procedure for drawing arrowheads.

Extension lines. Extension lines are also thin solid lines. They are drawn from the object to the terminal points of the dimension lines. Extension lines should begin 1/16 inches from the object and extend 1/8 in. beyond the dimension line. See Fig. 14.

Center lines. A center line is a thin broken line used to indicate the center of circles and other curved shapes as well as the central axis of a symmetrical object.

Cutting plane lines. Heavy broken lines drawn across objects to show where sectional views are taken are called cutting plane lines. You will learn more about sectional views shortly.

Section lines. Thin parallel lines drawn through the area in a sectional view to represent the cut material are called section lines.

Rounds and fillets

Unfinished surfaces of castings always have rounded edges where they intersect. This is done to prevent possible fractures at the point of inter-

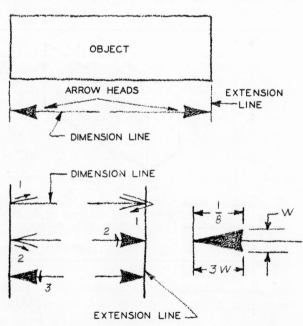

Fig. 14. Notice how dimension lines are drawn between extension lines.

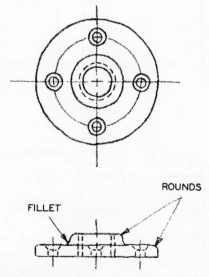

Fig. 15. Rounds and fillets are usually found on unfinished castings.

section. A rounded internal corner is referred to as a "fillet" and a rounded external corner as a "round." Notice in Fig. 15 how rounds and fillets are illustrated in a drawing.

Sectional views

In drawing some objects, it is difficult to show how the inside looks without using numerous hidden lines. To avoid this problem, a sectional view is often used. The sectional view assumes that a part of the object has been cut by an imaginary cutting plane and a section removed. See Fig. 16.

As you have seen in Fig. 10, the cutting plane or line is indicated on a drawing by either one of two types of cutting plane lines. One type consists of alternating long dashes and pairs of short dashes. See Figs. 10 and 17. The long dashes are made 3/4 in. to 1-1/2 in. in length and the short dashes approximately 1/8 in. in length with 1/16 in. spaces. The second type of cutting plane line is made of equal dashes about 1/4 in. long. See Figs. 10 and 17. The ends of both lines are bent 90° and end with arrowheads. The arrowheads should point to the direction of sight in which the object was viewed after the

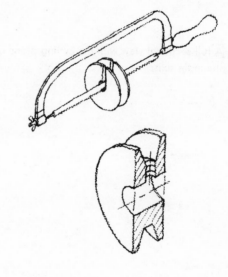

Fig. 16. A sectional view assumes that a part of the object has been removed to show the interior.

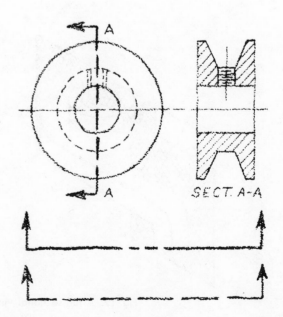

Fig. 17. Types of lines used to show the location of the cutting plane.

cut was made. Capital letters 3/16 in. to 3/8 in. high are placed behind the arrowhead. A notation is also placed under the section view, as SECTION A-A.

The main types of sectional views are known as *full section, half section, broken-section,* and *revolved section.*

Full section. When the cutting plane passes

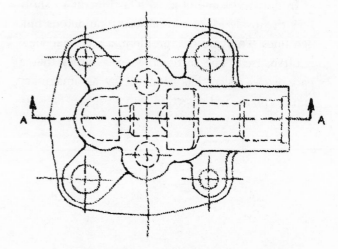

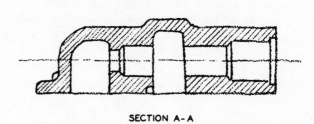

SECTION A-A

Fig. 18. A full sectional view with the cutting plane along the main axis.

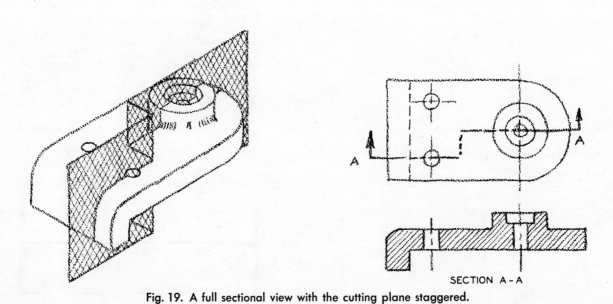

SECTION A-A

Fig. 19. A full sectional view with the cutting plane staggered.

entirely through the object, the result is a full section. The cutting plane may pass along the main axis of the object or it may be offset. See Figs. 18 and 19.

Half section. Two cutting planes passed at right angles to each other along the center lines or symmetrical axes result in a half-section. This permits the removal of one-quarter of the object as in Fig. 20.

Broken-out section. Where only a small portion of the interior needs to be shown, a broken-out section is used. The sectional area is outlined by a break line as in Fig. 21.

Revolved section. In order to show the true shape of objects such as bars, channels, spokes or ribs, a revolved section is used. In this case the cut section is turned through a 90° angle. See Fig. 22.

Equally spaced lines drawn at an angle of 45° are used to represent a sectional view. See Fig. 23. These lines should run in one direction except where there are two or more adjacent pieces. When this occurs, the lines are changed to run in

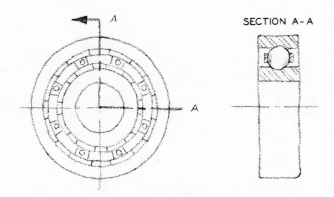

Fig. 20. A half sectional view results when one quarter of the object is removed.

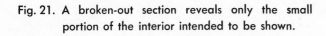

Fig. 21. A broken-out section reveals only the small portion of the interior intended to be shown.

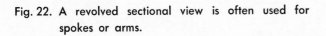

Fig. 22. A revolved sectional view is often used for spokes or arms.

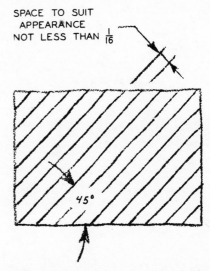

Fig. 23. Most sectional lines are shown at 45° angles.

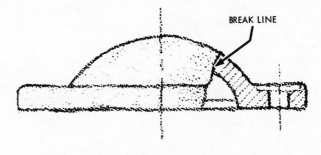

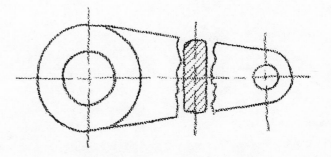

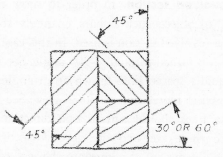

Fig. 24. By changing directions of the section lines, it is easier to identify different parts.

opposite directions so as to identify the pieces more easily. See Fig. 24.

Check your knowledge of this unit by completing the Self-Check for Unit 3 on pages 33 and 34.

Practice sketching the objects on Drawing Sheets 5, 6, 7 and 8, pages 35-38. Follow the instructions on the bottom of each sheet.

SELF-CHECK FOR UNIT 3

PART 1

DIRECTIONS: Circle the letter T if the statement is True or the letter F if the statement is False.
Self-Check answers may be found on page 172.

1. T F In a multiview drawing the usual practice is to show the right side view of an object.

2. T F A hole which appears as a circle in the front view is represented by hidden lines in the side view.

3. T F For some drawings only two views are required.

4. T F The weight of a visible outline should be the same as that of a dimension line.

5. T F A center line consists of equally-spaced dashes.

6. T F A rounded internal corner is known as a fillet.

7. T F Most objects require only three views for complete representation.

8. T F Arrowheads are used to indicate the ends of a dimension line.

9. T F Objects are positioned on a drawing sheet at right angles to the line of vision.

10. T F To position the object correctly, the object must be placed on the sheet so that the object is parallel to the plane of projection.

11. T F Objects are placed on the sheet so that the least number of hidden lines will have to be used.

12. T F In general, three types of lines are used on a drawing: heavy, medium, and light.

13. T F Visible lines are used to outline the visible sides of the object.

14. T F Arrowheads are three times as long as they are wide.

15. T F In sectioning objects two types of cutting plane lines may be used.

16. T F The arrowheads on a cutting plane line indicate the size of the section which is removed from the object.

17. T F The cutting plane line for a full section may be straight or offset.

18. T F In a half-section, the object is assumed to be cut in half.

PART 2

DIRECTIONS: Circle the letter which corresponds to the best answer to each of the following questions. *Self-Check answers may be found on page 172.*

1. A section where two cutting planes are passed at right angles to each other along the center lines or symmetrical axes is called
   a. a full section
   b. a revolved section
   c. a half section
   d. a broken-out section
   e. a cross-section

2. How many different types of cutting plane lines are there?
   a. one
   b. two
   c. three
   d. four

3. Arrowheads are used to
   a. dress up an otherwise dull drawing
   b. to indicate extension lines
   c. to indicate the end of a dimension line
   d. to show where a sectional view is taken

4. Arrowheads are
   a. as long as they are wide
   b. twice as long as they are wide
   c. varying sizes, depending on the size of the object being dimensioned
   d. three times as long as they are wide

5. Which of the following is NOT a thin weight line?

   a. a hidden line
   b. a center line
   c. a section line
   d. an extension line

6. Which of the following views is usually NOT used on a drawing?
   a. a top view
   b. a left side view
   c. a right side view
   d. a front view

7. Which of the following statements is most true?
   a. An object always has three views
   b. An object should have five views for correct understanding of the object
   c. An object will have only as many views as is necessary to the understanding of the object
   d. Many times only one view is needed

   Score_____

A Sketch a side view.

B Sketch front and top views.

C Sketch front and side views.

D Sketch the required views.

E Sketch the required views.

F Sketch the required views.

Make freehand sketches of the required views for each of the above.

TITLE		DRAWING NO.
NAME	GRADE	

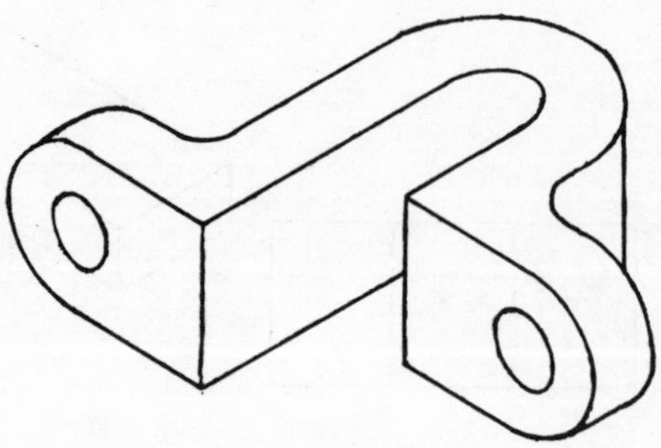

A

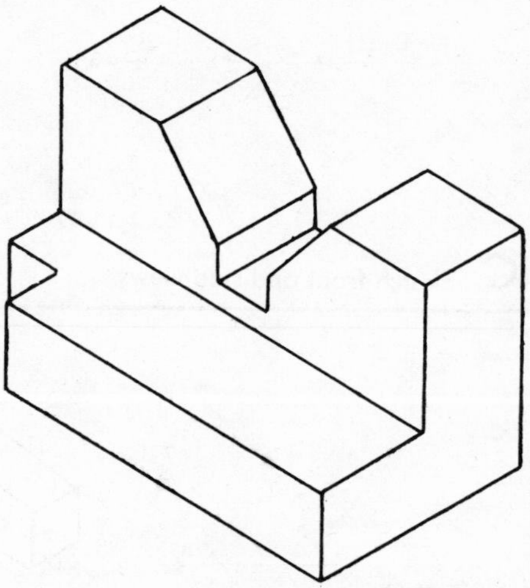

B

	TITLE		DRAWING NO.
Make three-view sketches of the objects shown above.	NAME		GRADE

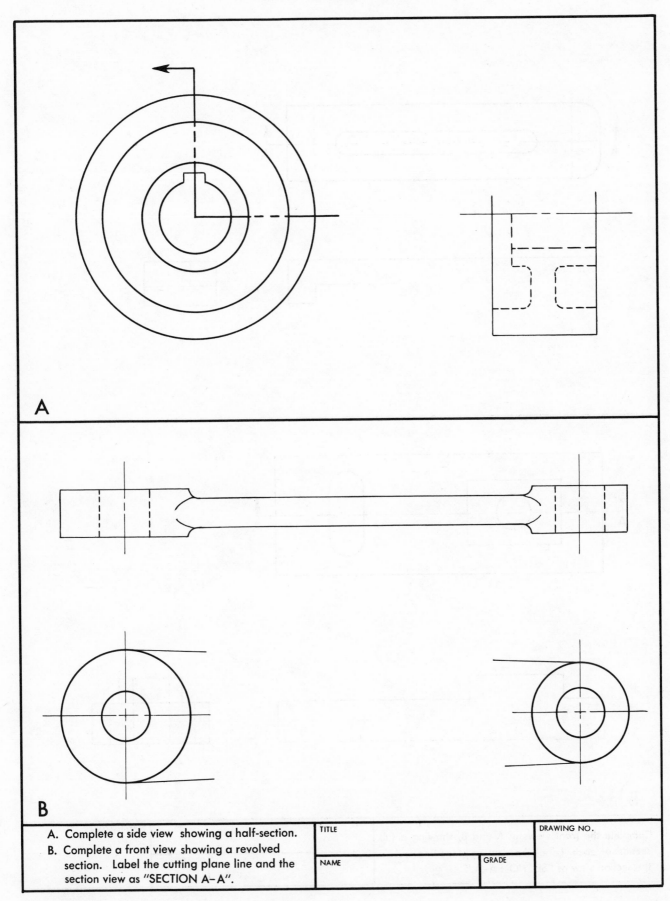

A

B

A. Complete a side view showing a half-section.
B. Complete a front view showing a revolved
   section. Label the cutting plane line and the
   section view as "SECTION A–A".

TITLE		DRAWING NO.
NAME	GRADE	

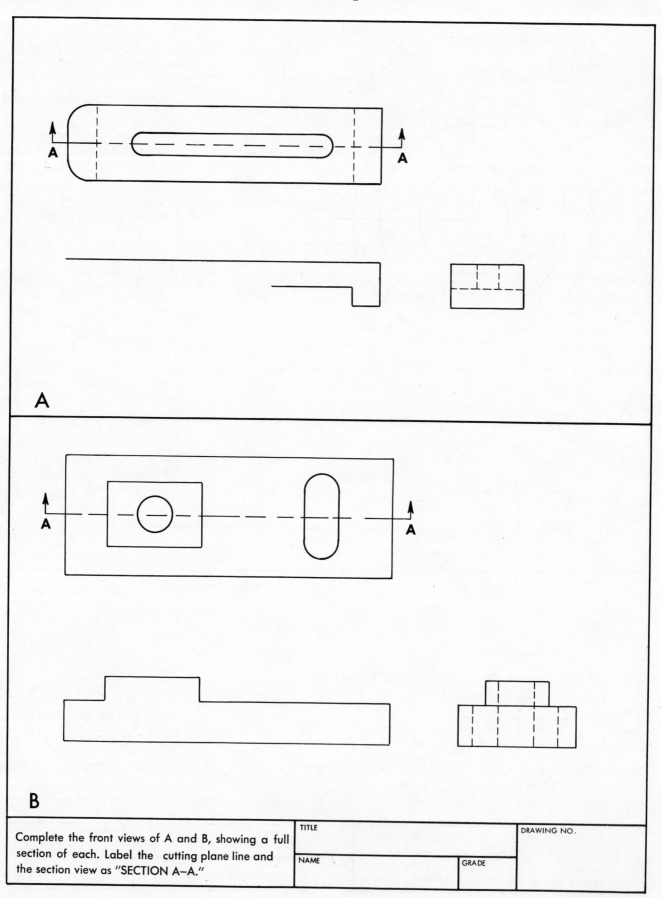

A

B

Complete the front views of A and B, showing a full section of each. Label the cutting plane line and the section view as "SECTION A–A."

TITLE		DRAWING NO.
NAME	GRADE	

Lettering

In addition to showing the various views of an object, a drawing also contains information in the form of words and numerals. The process of putting this information on a drawing is known as lettering. Since lettered information is an essential part of a drawing, the letters and numerals must be easy to read, pleasing in appearance, and uniform in style. Poor lettering can often detract from a good drawing.

For most drawings, the general practice is to use uppercase (capital), single-stroke Gothic letters. Lower case (small) letters are used mostly for maps, architectural, and structural drawings. Incidentally, single-stroke simply means that the width of the lines which form the letters does not vary. It does not mean that letters are begun and

completed in one single stroke. The letters can be drawn with either a vertical or slanted style. See Figs. 1 and 2. Try out both styles and then concentrate on the one that appeals to you more and that you can make more easily. Both styles are acceptable, although the vertical is more commonly used.

Forming letters and numbers

There are two things you must keep in mind to form letters and numbers correctly: (1) letters and numbers vary in width and (2) the strokes in making letters and numbers must follow certain directions.

If you study Fig. 3, you will see the variation

Fig. 1. Vertical strokes may be used in making letters and numbers.

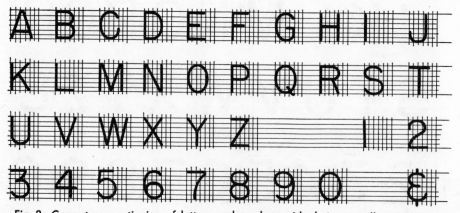

ABCDEFGHI
JKLMNOPQR
STUVWXYZ&
1234567890

Fig. 2. Slanted strokes may be used in making letters and numbers.

in widths of letters. The difference in width gives the letters more pleasing proportions. Thus some letters are almost as wide as they are high, while others are slightly narrower. To give you a better idea of the relationship between width and height, the letters are reproduced in blocks which are six squares high and six squares wide. Notice that most letters are approximately two-thirds as wide as they are high. Letters B, C, D, E, F, G, etc., are four squares wide. Others, such as A, O, Q, R, and V are four and one-half squares wide, whereas M is five squares wide. The widest letter of the alphabet is W, which is almost six and one-half squares

wide. Fig. 3 also shows the correct proportions for numbers.

Now look at Figs. 1 and 2 again, and see how the strokes are made for each letter and number. Do not try to memorize all these strokes. When you begin working on the drawing sheets for this unit, use Figs. 1 and 2 as guides. After you have made these letters and numbers a few times, you will find yourself automatically making the strokes correctly.

Spacing letters and words

To secure a pleasing appearance, the areas be-

Fig. 3. Correct proportioning of letters and numbers aids their over-all appearance.

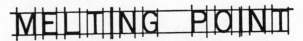

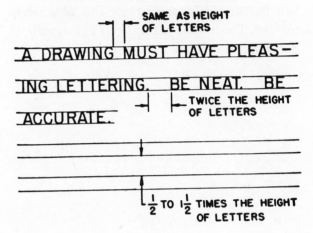

Fig. 4. Correct spacing between letters, words, and sentences is necessary for an even appearance.

tween each letter must appear to be equal. Due to the shape of various letters, the appearance of equal areas cannot always be obtained by simply setting the letters a uniform distance apart. The usual practice is to judge the spacing by eye.

The distance between words should be equal to the height of the letters. See Fig. 4. Sentences are spaced at a distance equal to about twice the distance between words. When several lines are required, the spacing between them may vary from one-half to one-and-one-half times the height of the letters.

Fig. 5. Standard heights of letters.

Height of letters and words

In general, letters for notes and dimensions are made 1/8 in. high. Titles usually are 3/16 in. to 1/4 in. high and letters to indicate sections are 5/16 in. high. See Fig. 5.

Height of numerals

Whole numbers and decimals should be made the same height as the letters. The height of each number in a fraction should be approximately

Fig. 6. Standard heights of numbers.

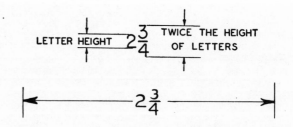

three-fourths the height of the whole number. The full height of the fraction including the fraction bar should be about twice that of a whole number. The fraction bar is always in line with the dimension. See Fig. 6.

Check your knowledge of this unit by completing the Self-Check for Unit 4 on page 43.

Practice lettering on Drawing Sheets 9, 10, and 11, pages 44-46.

SELF-CHECK FOR UNIT 4

DIRECTIONS: Circle the letter T if the statement is True or the letter F if the statement is False.

Self-Check answers may be found on page 172.

1. T F All letters are of identical width.

2. T F Notes and sizes on drawings should be written in longhand.

3. T F Areas between letters should appear to be equal.

4. T F The process of lettering is simplified if letters are formed with specific strokes.

5. T F Vertical lettering is preferable to the slanted.

6. T F Spacing between lines may vary from one-half to one-and-one-half times the height of the letters.

7. T F Letters for titles are usually the same height as those used for notes.

8. T F In lettering a drawing, the widest letter is M.

9. T F In lettering, the same amount of space must be allowed between each letter.

10. T F The distance between words should be equal to the height of the letters.

11. T F Whole numbers are made the same height as the letters.

12. T F The fraction bar of a fraction should be slightly above the dimension line.

VERTICAL

A
B
C
D
E
F
G
H
I
J
K
L
M
N
O
P
Q
R
S
T
U
V
W
X
Y
Z

INCLINED

A
B
C
D
E
F
G
H
I
J
K
L
M
N
O
P
Q
R
S
T
U
V
W
X
Y
Z

Complete each line of lettering.

TITLE		DRAWING NO.
NAME	GRADE	

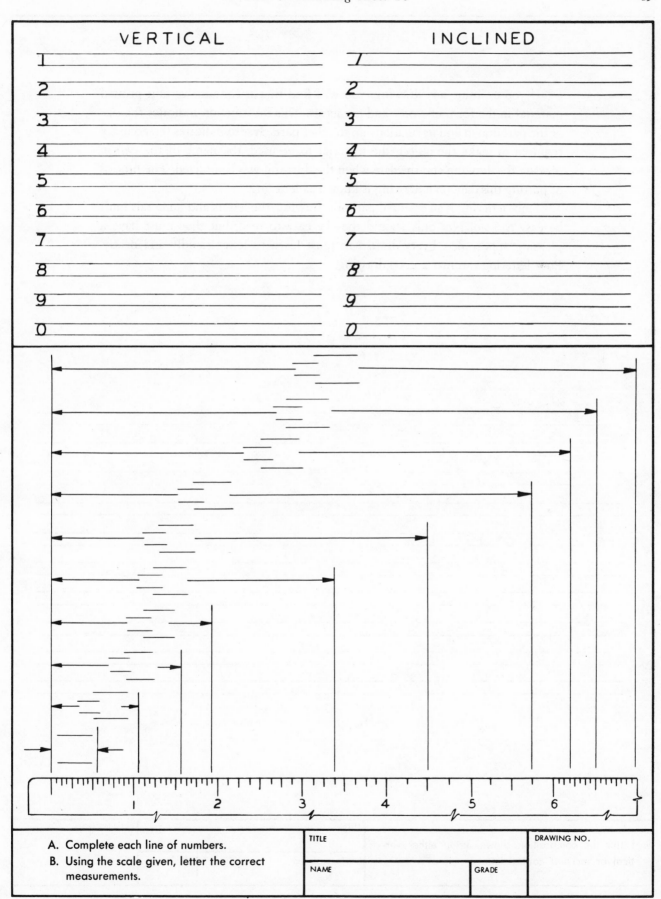

VERTICAL INCLINED

1
2
3
4
5
6
7
8
9
0

A. Complete each line of numbers.
B. Using the scale given, letter the correct measurements.

TITLE

NAME

GRADE

DRAWING NO.

If you examine any drawing, you will find that it contains certain printed material in the form of words and numerals. This information indicates the size of the part drawn and its relationship to other parts. It also indicates the accuracy required to make the object, the material to be used, the name of the person making the drawing and the date when the drawing was completed. The process of putting this data on a drawing is known as lettering.

Since lettering is an essential part of a drawing, the letters and numerals must be easy to read. Not only should they be easy to read, but they must have a pleasing appearance. Every draftsman must learn to letter rapidly and neatly. Poor lettering can mar a good drawing.

Letter the composition shown, using either vertical or inclined capital letters.	TITLE		DRAWING NO.
	NAME	GRADE	

Dimensioning

Drawings and sketches are dimensioned in inches, feet and inches, or decimals. The symbol for the inch is ″ and for the foot ′. The inch symbol is omitted if all the dimensions are in inches. When dimensions are given in feet and inches, both symbols are used with a hyphen between them as 8′-7½″.

Angular dimensions are expressed in degrees °, minutes ′, and seconds ″

Dimensioning a drawing

The sizes of an object are indicated by numbers placed within dimension lines. Dimension lines are thin, solid lines terminated with arrowheads. See Unit 3, Fig. 10. They are drawn between extension lines and are at least 3/8 in. from the object. See Fig. 1. The extension lines are drawn so that they begin about 1/16 in. from the object.

Dimension lines should be aligned whenever possible and grouped uniformly as shown in Fig. 2.

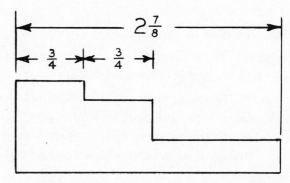

Fig. 2. Group dimensions so they will produce an orderly appearance.

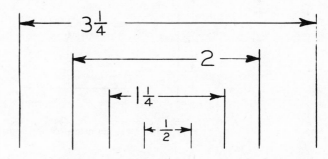

Fig. 3. Stagger the numerals when parallel dimension lines are required.

All parallel dimension lines should never be less than 1/4 in. apart. If several parallel dimension lines are necessary, the numerals should be staggered for easier readability. See Fig. 3.

Dimensions are usually placed so as to read from the bottom of the sheet. See Fig. 4.

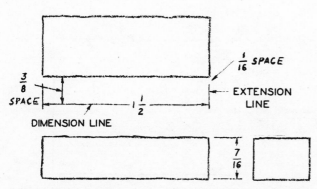

Fig. 1. Correct placement of dimension and extension lines is important.

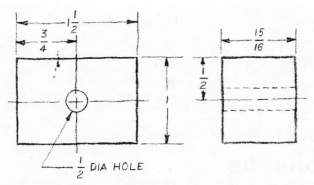

Fig. 4. Arrange dimensions to read from the bottom of the sheet.

Placement of dimensions

Here are a few basic rules which are considered standard practice in dimensioning a drawing:

1. The most important dimensions should be located at the principal view of a part. It is usually this view that most completely shows the essential contour characteristics of the piece. See Fig. 5.

2. Dimensions should not originate at or terminate on hidden lines.

3. Arrowheads should be approximately 1/8 in. on small drawings and up to 3/16 in. on large drawings. The width of the arrowhead should be about one-third its length. See Fig. 6.

4. Do not repeat dimensions.

5. Locate dimension lines so they will not cross extension lines or other dimension lines. See Fig. 7.

6. Keep dimensions off the views. See Fig. 8.

7. Place the larger dimensions the farthest away from the views. See Fig. 7.

8. Notice in Fig. 9 how to arrange dimensions for small areas.

Dimensioning circles, arcs and angles

Locate centers for holes and circles by means of center lines. By extending center lines beyond the

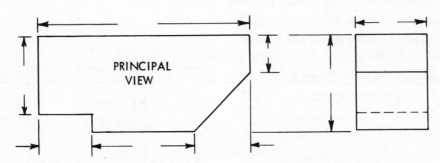

Fig. 5. The most important dimensions should be located at the principal view.

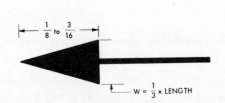

Fig. 6. How arrowheads should be made.

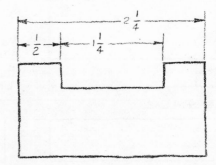

Fig. 7. Keep dimension lines from crossing extension lines.

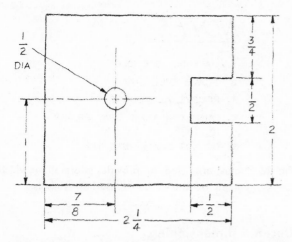

Fig. 8. Keep dimension lines off the view, if possible.

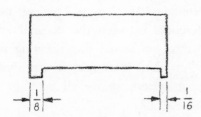

Fig. 9. How to arrange dimensions in small places.

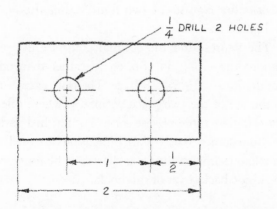

Fig. 10. When dimensioning the position of circles, always indicate the distance between center lines.

view, they can often be used in place of extension lines. See Fig. 10.

Always give the diameter of a circle and not its radius. On small circles, place the dimension on the outside with the letters DIA. On large circles, dimensions may be placed either on the inside or outside of the circle as in Fig. 11. Locate the dimension of holes and circles on the view where

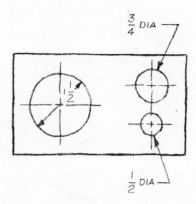

Fig. 11. Dimensions of large circles can be either inside or outside the circle.

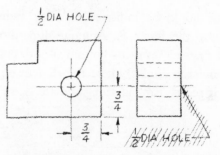

Fig. 12. Do not dimension circles where they appear as hidden lines.

they appear as circles and not on the view where they are represented by hidden lines. See Fig. 12.

When equally spaced holes are located around a circle, give the diameter of the circle across the circular center line and the number and size of the holes in a note. See Fig. 13. If the holes are unequally spaced, show them as in Fig. 14.

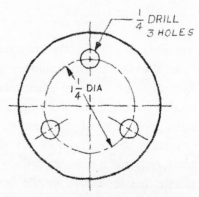

Fig. 13. To dimension equally spaced identical circles, use only one dimension and indicate that it applies to all circles.

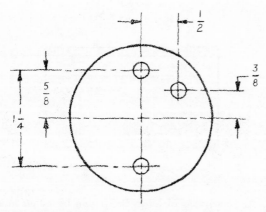

Fig. 14. Full dimensions must be given for unequally spaced holes.

Arcs should be indicated by their radius followed by the letter R. Angular dimensions are placed to read from the bottom of a drawing except on large angles where the dimension sometimes is placed to read along the arc. See Fig. 15.

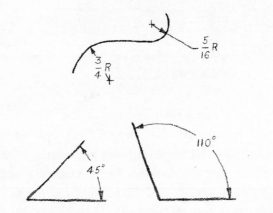

Fig. 15. Dimensions of arcs and angles are usually read from the bottom of the sheet.

Notes

Notes are used to explain the material from which the part is to be made, the number of pieces required, type of finish and any other data which may be needed to make the drawing complete. Notes are also used to avoid repetition of dimensions. For example, where a number of equal radii or holes occur, only one is dimensioned and a note used to indicate that the others are the same. The important point to remember when including notes is to make them brief, but specific. See Fig. 16.

NOTES

1 MATERIAL SAE 1020
2 SCALE FULL SIZE
3 REMOVE ALL BURRS
4 PAINT ONE COAT GRAY PRIMER IN SHOP
5 DO NOT SCALE DRAWING

Fig. 16. Notes are used to provide information which cannot be shown by measuring symbols.

Decimal dimensioning

Most industries today use the decimal system of dimensioning because it provides more exacting control of manufactured parts. There are no significant differences between the decimal and fractional methods in actual dimensioning practices. Whereas one designates sizes in fractional units, the decimal system shows all sizes in decimal values. See Fig. 17.

The decimal dimensioning system is based on the use of two-place decimals; that is, decimal dimensioning consists of two figures after the decimal point.

The figures after the decimal point are in even hundredths—.04, .34, .86, etc.,—rather than odd hundredths—.03, .35, .87, etc. The only exception to the use of the two-place decimal is where tolerances require more precise control of finished parts. In this instance 3- or 4-place dimensions are used. A fraction-decimal conversion chart may be found on the inside back cover of this book.

Tolerances

A draftsman will frequently specify an allowable margin within which a measurement can vary. This allowable error is referred to as "tolerance."

The amount of tolerance which is permitted depends upon the accuracy or tightness required for parts to function. These tolerances are usually expressed in decimals, although where the values are not too critical they may be shown in fractions. See Fig. 18.

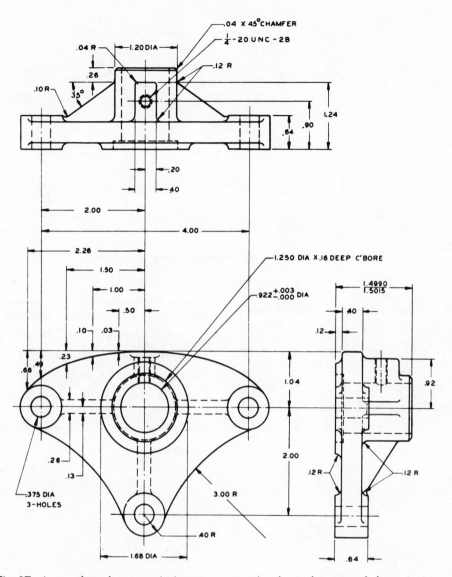

Fig. 17. A complete drawing which incorporates the decimal system of dimensioning.

Fig. 18. On some drawings it is necessary to specify the tolerance that a measurement can vary.

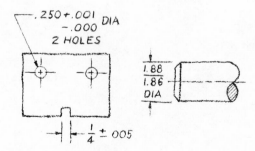

When it is necessary to show the amount of variation for any given dimension, the practice is to:

1. Specify two tolerance numerals, one plus and one minus, if the plus variation differs from the minus variation. See Fig. 19 (Top left).

2. If the plus variation is equal to the minus

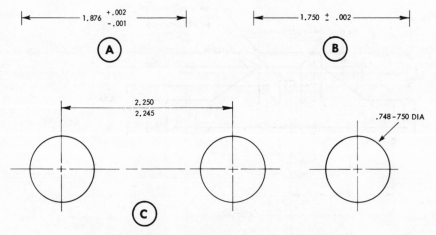

Fig. 19. Method of showing variations of dimensions.

variation a combined plus and minus size is used. See Fig. 19 (Top right).

3. Place the high limit above the low limit where dimensions are given directly. When dimensions are given in note form, the low limit precedes the high limit. See Fig. 19 (Bottom).

Check your knowledge of this unit by completing the Self-Check for Unit 5 on pages 53 and 54.

After you are thoroughly familiar with dimensioning procedures, apply your knowledge by drawing and dimensioning the objects on Drawing Sheets 12, 13, 14, 15, 16 and 17, pages 55-60.

SELF-CHECK FOR UNIT 5

PART 1

DIRECTIONS: Circle the letter T if the statement is True or the letter F if the statement is False. *Self-Check answers may be found on page 172.*

1. T F When the dimension includes a fraction, the lower number of the fraction is in line with the dimension line.

2. T F Dimensions are read from wherever it is most convenient to place them.

3. T F If several identical circles are shown, a note may be used with just one circle indicating the size of all the circles.

4. T F When dimensions are staggered, the larger dimensions are placed closer to the object than the smaller dimensions.

5. T F When dimensioning the position of a circle on an object, the dimension line is placed from the center line of the circle to the end of the object.

6. T F The inch mark should be included in all dimensions.

7. T F Dimension lines should be kept at least ⅜ in. from the views.

8. T F In dimensioning circles, the radius should be given.

9. T F The sizes of circles may be placed either on the inside or outside of the circle.

10. T F Tolerance is used to show how much a size can vary from the required dimension.

PART 2

DIRECTIONS: Circle the letter which corresponds to the best answer to each of the following questions. *Self-Check answers may be found on page 172.*

1. Dimensions are usually placed so as to read
   a. from the side of the sheet
   b. from whatever position is easiest to place the dimension
   c. from the bottom of the sheet
   d. from the top of the sheet

2. Which of the following rules is NOT correct?
   a. Place dimension lines between views
   b. Never stagger dimensions
   c. Keep dimensions off views
   d. Do not repeat dimensions

3. Which of the following statements is correct?
   a. The radius of a circle should always be given, never its diameter
   b. When equally-spaced holes of the same size are dimensioned, each hole must be dimensioned separately
   c. Notes are used to explain the material from which the part is to be made, the number of pieces required, type of finish and any other data necessary to complete the drawing
   d. Tolerances are usually expressed in units of 1/16 in.

4. How far should parallel dimension lines be drawn apart from each other?
   a. ¼ inch
   b. ⅜ inch
   c. ½ inch
   d. ⅛ inch

5. Which of these directions is *incorrect?*
   a. Do not repeat dimensions
   b. Place the most important dimensions on the principal view
   c. Place the largest dimensions closest to the view
   d. Avoid having dimension lines cross extension lines

6. Which of these statements is true?
   a. Principal dimensions are located within a view
   b. Dimensions of large circle are never placed inside the circle

   c. Where they appear as hidden lines, circles should not be dimensioned
   d. Arrowheads are made so their width is equal to their length

7. The amount of tolerance which is specified on a drawing is governed by
   a. size of the part
   b. shape of the part
   c. design of the part
   d. accuracy or tightness required for parts to function

8. In decimal dimensioning, which of these statements is true?
   a. A three-place decimal is usually required
   b. Most sizes are expressed as two-place decimals
   c. Decimals should be shown in odd rather than in even hundredths
   d. Decimals dimensions are used only to show tolerances

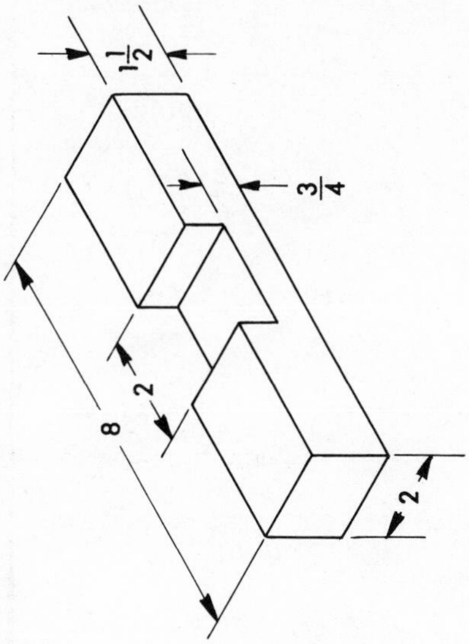

$1\frac{1}{2}$

$\frac{3}{4}$

8

2

2

DRAWING NO.

GRADE

TITLE

NAME

Sketch a three-view drawing and dimension completely. Make sketch ½ size.

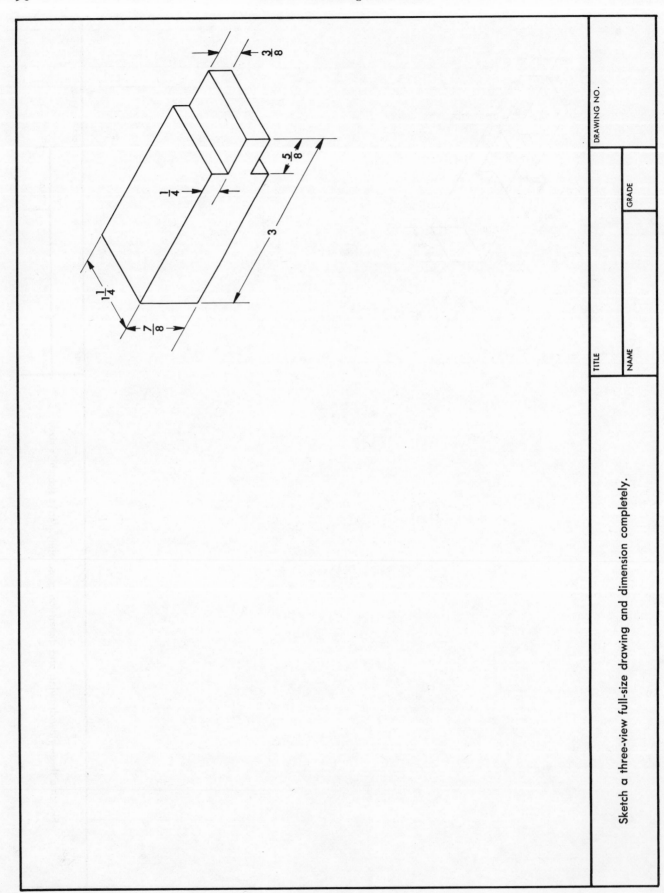

DRAWING NO.

GRADE

TITLE

NAME

Sketch a three-view full-size drawing and dimension completely.

TITLE		DRAWING NO.
NAME	GRADE	

Sketch a three-view drawing and dimension completely. Make drawing ½ size.

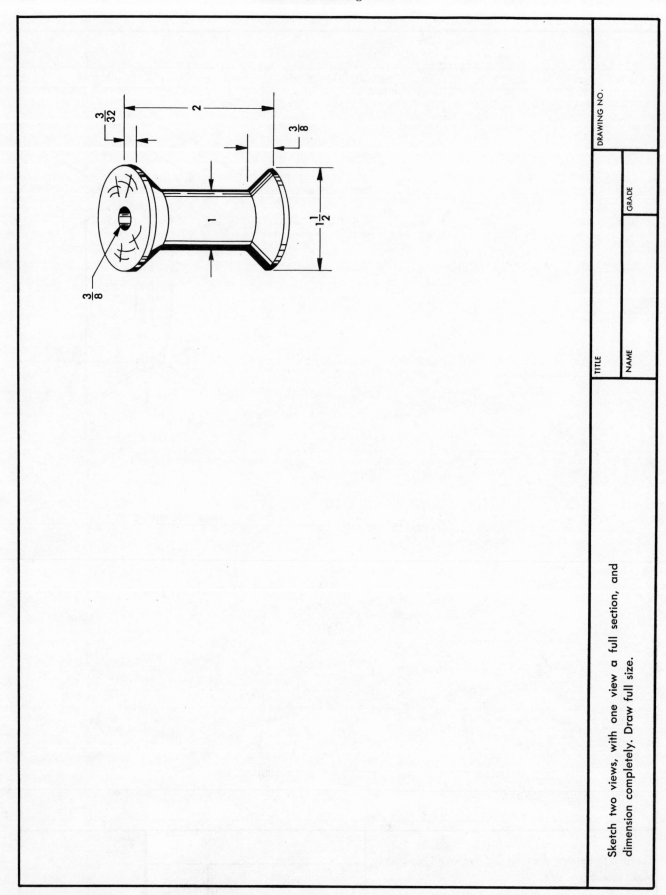

DRAWING NO.

GRADE

TITLE

NAME

Sketch two views, with one view a full section, and dimension completely. Draw full size.

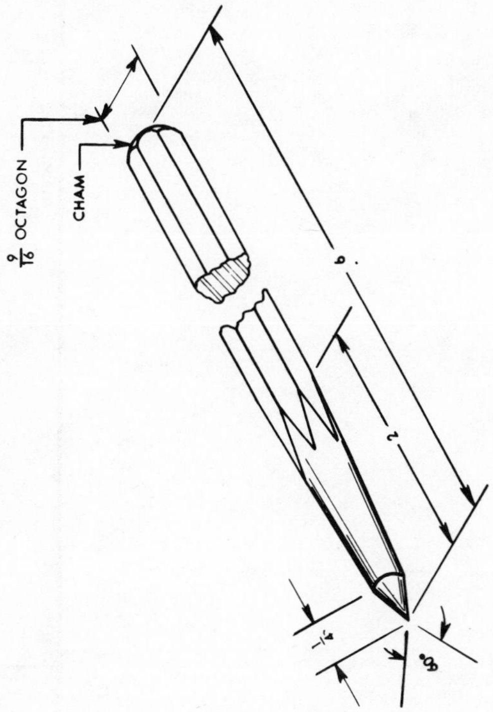

OCTAGON $\frac{9}{16}$

CHAM

TITLE

NAME

GRADE

DRAWING NO.

Sketch two views showing a revolved section in octagon portion and dimension completely. Make drawing full size.

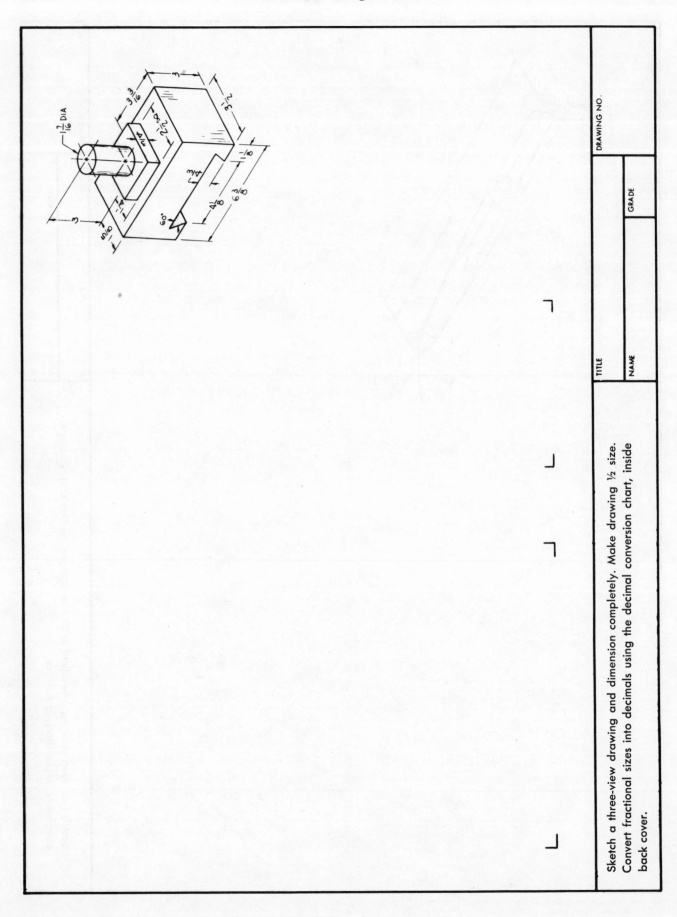

DRAWING NO.

GRADE

TITLE

NAME

Sketch a three-view drawing and dimension completely. Make drawing ½ size. Convert fractional sizes into decimals using the decimal conversion chart, inside back cover.

Showing Fasteners on Working Drawings

In making a drawing there will be occasions when you will have to show how parts are to be assembled. For example, the object which you are drawing may require that the pieces be bolted, riveted, or welded. If the drawing is to be an accurate representation of what is to be done, then you will need to indicate what fasteners are to be used. In this unit you will learn how fasteners are drawn as well as how they should be specified.

Working drawings

Working drawings, which are often referred to as production drawings, consist of *detail drawings* and *assembly drawings*.

Detail drawings. A drawing containing all of the information necessary for the actual construction of an individual part of the object is called a detail drawing. It may have one, two, three, or more views as you learned in the previous unit. In addition to dimensions, it must show the material from which the part is to be made. See Fig. 1A, the detail drawing on the following page.

If the product is small and consists of only a few parts, the details of each piece are sometimes grouped together on one large sheet. Occasionally these details are placed with the assembly drawing. In general, however, the practice is to draw each detail on a separate sheet.

Moreover, many industries prepare separate drawings to cover each specific manufacturing process that the object undergoes. Thus there may be a pattern drawing, a forging drawing, several machining drawings, a welding drawing, and a stamping drawing.

Assembly drawings. A drawing which presents the object as it appears after all of the parts, drawn separately in detail drawings, are fitted together is called an assembly drawing. Although there are several types of assembly drawings, we are going to concentrate on what is known as a general assembly drawing. See Fig. 1B, the assembly drawing.

In a general assembly drawing, only one or two views are shown: a principal view and, if necessary, a sectional view. If the mechanism is not very complicated, only a single sectional view may be used.

Only principal dimensions such as over-all height, width, or length are shown. Detailed dimensions of individual parts are not shown since they may be obtained from the detail drawings.

On a general assembly drawing, each part is identified by a leader line touching the outline of the part and terminating with a circle approximately 3/8 in. in diameter. These circles, called "balloons," contain numbers. The numbers identify the pieces in a "Parts List." The parts list is a chart which names each piece and tells how many are required. See the assembly drawing on page 63. The parts list is located on the assembly drawing or is made on a separate sheet.

It will prove helpful to remember the following rule: when objects are relatively simple, both the details and the assembly may be included in one

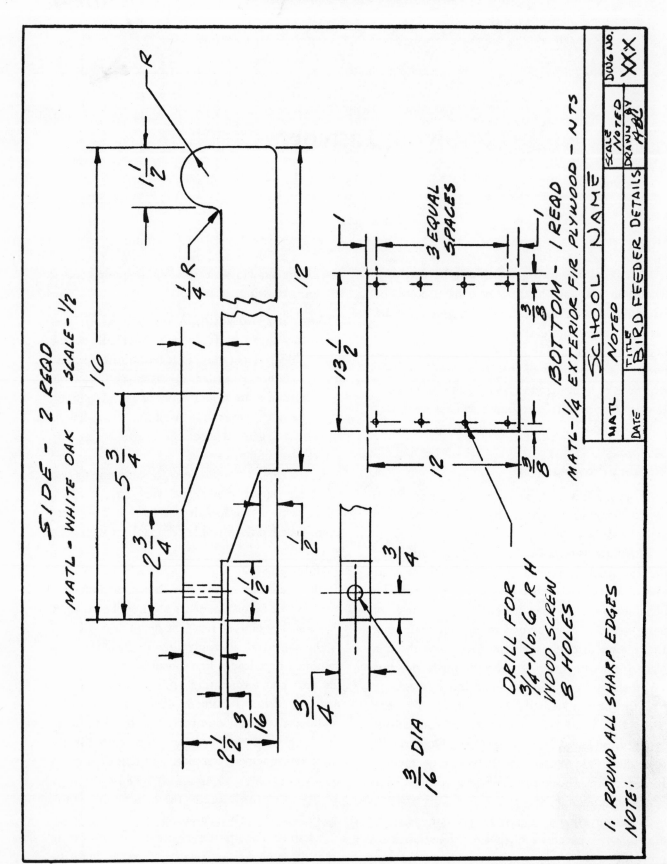

Fig. 1A. A detail drawing provides all of the information required to construct the part.

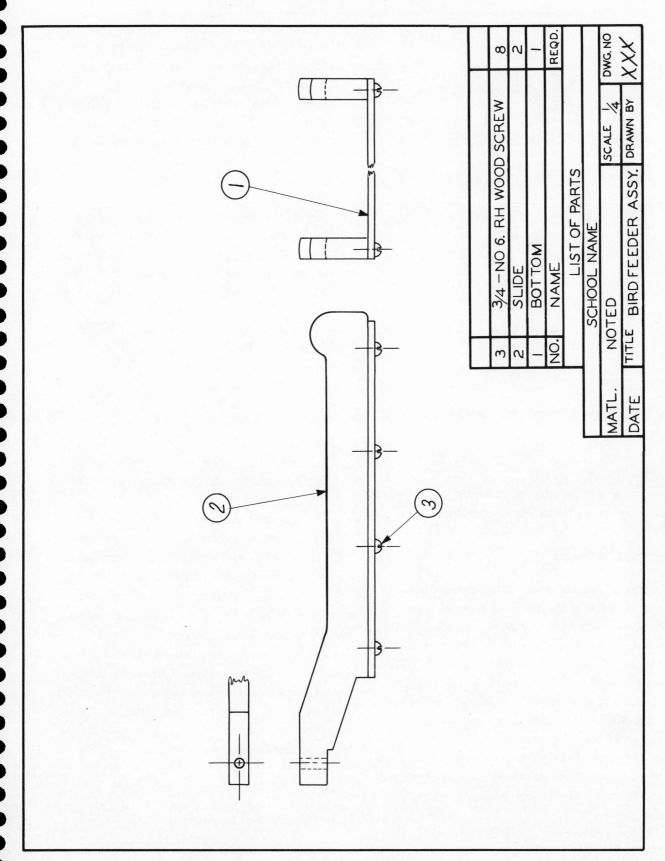

Fig. 1B. An assembly drawing shows how parts are fastened together.

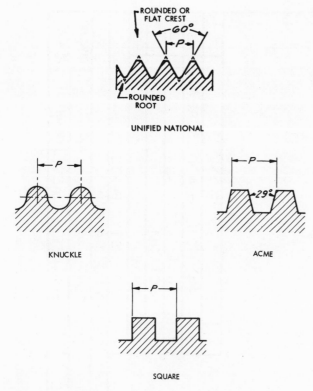

Fig. 2. The types of threads shown are most commonly used.

Coarse thread, designated Unified National Coarse (UNC). This thread is used on screws, bolts, and nuts for general application requiring rapid assembly or disassembly.

Fine thread, designated as Unified National Fine (UNF). This thread has a greater number of threads per inch for each diameter and has more holding power than the coarse thread, especially where parts are subject to considerable vibrations.

Extra-Fine thread, designated as Unified National Extra-Fine (UNEF). This thread is used for threaded parts which require fine adjustment such as on bearing retaining nuts and on thin nuts where maximum thread engagement is needed.

Thread classes. Threads are made with different degrees of fit. Fit refers to the allowance (looseness or tightness) between the mating screw and nut. The Unified National Screw Thread Series has three classes of fit. These are:

Class 1 possesses the largest allowance and is used where rapid assembly of parts is required and looseness or play is not objectionable.

Class 2 is used on the bulk of standard screws, bolts and nuts.

Class 3 is used on fasteners where accuracy is very important and no looseness is permitted.

In addition to the classification number, the letter A or B is included to designate an external or internal thread. A thread on the external surface of a cylinder or cone is called an *external thread.* A thread on the internal surface of a hollow cylinder or cone is called an *internal thread.* Thus 3A represents a Class 3 fit on an external thread. Similarly, 3B indicates the same fit on an internal thread. Most threaded fasteners are referred to in terms of type, coarseness, and class. See the specification for the thread in Fig. 3 (Bottom).

drawing. For most objects, however, the details and assembly are shown on separate sheets.

Threaded fasteners

Threaded fasteners are those whose holding power is obtained by means of threads. Some of the more common varieties of threads are the Unified National, the Square, the Acme, and the Knuckle. See Fig. 2. The Unified National, is the most frequently used type of thread. The Unified National thread is essentially the former American National type. The differences are primarily in the degree of fit tolerances and in the shape of the root and crest of the thread. This thread was developed to permit greater interchange of threaded parts between the United States, Canada, and Great Britain.

Thread coarseness. Threads are made in different degrees of coarseness. Those with which you will be mostly concerned are:

How threads are represented

Threads are drawn by using either the *schematic* or *simplified* representation. To produce a *sche-*

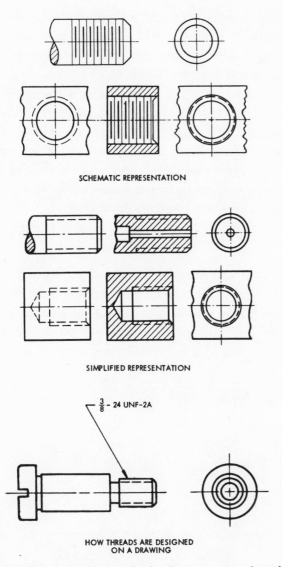

SCHEMATIC REPRESENTATION

SIMPLIFIED REPRESENTATION

$\frac{3}{8}$ - 24 UNF-2A

HOW THREADS ARE DESIGNED
ON A DRAWING

Fig. 3. There are three methods of representing threads on a drawing.

Types of threaded fasteners

Bolts. A fastener having a head on one end and a stem that fits through an opening in two or more parts is referred to as a bolt. The opposite end of the bolt is threaded to receive a nut which holds

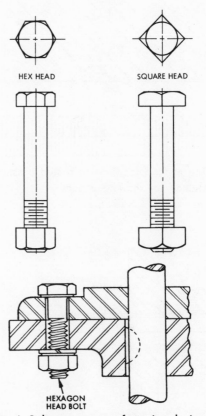

HEX HEAD   SQUARE HEAD

HEXAGON
HEAD BOLT

Fig. 4. Bolts are common fastening devices.

matic thread symbol, Fig. 3 (Top), crest lines are uniformly drawn to show threads. The line spacing need not be the actual pitch distance, however the lines should not be closer than 1/16″. The root lines are centered between the crests and terminated a short distance from the outside diameter of the thread.

The *simplified* representation, Fig. 3 (Center), is made by simply drawing two parallel invisible lines to designate the root of the thread. Fig. 3 (Bottom) illustrates how threads are specified, or dimensioned, on a drawing.

the parts together. See Fig. 4. On a drawing, the specification of bolts should show the diameter, number of threads per inch, type of thread, coarseness, class of fit, length and type of head. For example, a bolt would be designated as:

$$\text{3/8} - 16 \text{ UNC} - 2A \times 2\text{1/2 HEX HD BOLT}$$

Studs. A rod threaded on both ends is called a stud or stud bolt. It is used when regular bolts are not suitable, especially on parts that must be removed frequently, such as cylinder heads. One end of the stud is screwed into a threaded hole, and the other end fits into the removable piece of the struc-

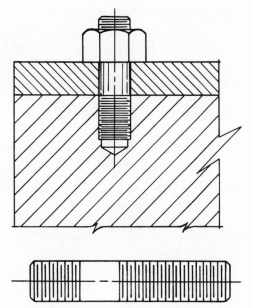

Fig. 5. The method for showing a stud is similar to the method for showing a bolt.

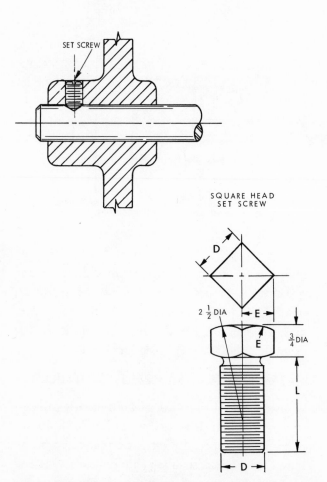

SET SCREW

SQUARE HEAD
SET SCREW

Fig. 6. This set screw prevents the shaft from turning in the hub.

ture. A nut is used on the end projecting thru the removable piece to hold the parts together. See Fig. 5. Specifications for studs are listed in the same way bolts are listed.

Set screws. The function of a set screw is to prevent rotary motion between two parts, such as the hub of a pulley and a shaft. See Fig. 6. The set screw is driven into one part so that its point bears firmly against another part. Set screws are either headless or have a square head. On a drawing, a set screw is usually designated as:

¼ — 20 UNC — 2A x ½ SQ HD SET
SCR CUP POINT

Cap and machine screws. A cap screw passes through a clearance hole in one part of an object and screws into a threaded hole in another part. See Fig. 7. Machine screws are similar to cap

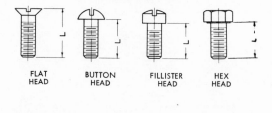

| FLAT HEAD | BUTTON HEAD | FILLISTER HEAD | HEX HEAD |

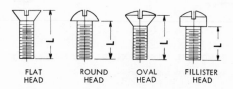

| FLAT HEAD | ROUND HEAD | OVAL HEAD | FILLISTER HEAD |

Fig. 7. Cap screws (Top) and machine screws (Bottom) have a wide variety of heads.

screws except that they are smaller and are used chiefly on work having thin sections. Cap and machine screws are specified as:

No. 10-24 UNC — 2A x ⅜ RD HD MACH SCR
½ — 13 UNC — 2A x 1 FLAT HD CAP SCR

Wood screws. Wood screws are made of steel, brass, bronze, and aluminum. They are available in three types of heads: flat, oval, and round. See Fig. 8. Sizes of screws are specified according to

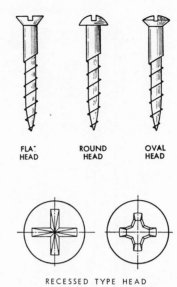

Fig. 8. Wood screws, like self-tapping screws, specify wire gage number instead of diameter.

the length, wire gage number (diameter), shape of head, and material. A typical specification would be:

3/4 — No. 6 FH WOOD SCR STEEL

Self-tapping screws. Screws used to fasten sheet metal parts or to join light metal, wood, or plastic are called self-tapping screws. These screws form their own threads in the material as they are turned. See Fig. 9. On a drawing, specifications for self-tapping screws include the length, wire gage number (diameter), and type.

3/8 — No. 4 Type A TAPPING SCR

Types of non-threaded fasteners

Nails. There are many different kinds of nails, such as the common, box, finishing, casing, and brad. See Fig. 10. Sizes of nails are designated by the term "penny" (d) with a prefix number such as 4d, 10d. The term "penny" refers to the weight of the nail per thousand in quantity. A 6d nail means that the nails weigh six pounds per thousand. This weight has a direct relationship to the length and diameter. Brads are the smallest types of nails; their sizes are indicated by the length in inches and diameter by the gage number of the wire. On a drawing, nails and brads are shown as:

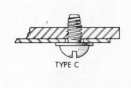

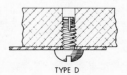

Fig. 9. Self-tapping screws are primarily used in sheet metal assembly.

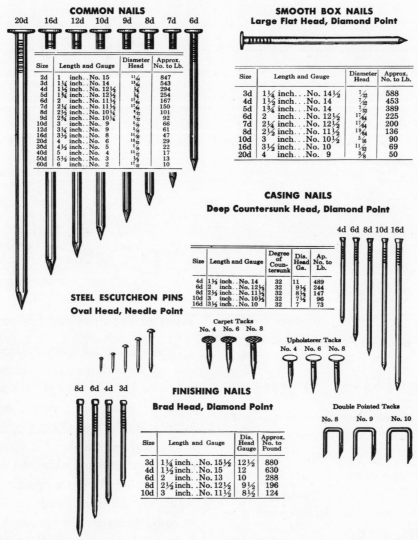

COMMON NAILS

Size	Length and Gauge	Diameter Head	Approx. No. to Lb.
2d	1 inch..No. 15	11/64	847
3d	1¼ inch..No. 14	13/64	543
4d	1½ inch..No. 12½	¼	294
5d	1¾ inch..No. 12½	¼	254
6d	2 inch..No. 11½	17/64	167
7d	2¼ inch..No. 11½	17/64	150
8d	2½ inch..No. 10¼	9/32	101
9d	2¾ inch..No. 10¼	9/32	92
10d	3 inch..No. 9	5/16	66
12d	3¼ inch..No. 9	5/16	61
16d	3½ inch..No. 8	11/32	47
20d	4 inch..No. 6	13/32	29
30d	4½ inch..No. 5	7/16	22
40d	5 inch..No. 4	15/32	17
50d	5½ inch..No. 3	½	13
60d	6 inch..No. 2	17/32	10

SMOOTH BOX NAILS
Large Flat Head, Diamond Point

Size	Length and Gauge	Diameter Head	Approx. No. to Lb.
3d	1¼ inch...No. 14½	7/32	588
4d	1½ inch...No. 14	7/32	453
5d	1¾ inch...No. 14	7/32	389
6d	2 inch...No. 12½	17/64	225
7d	2¼ inch...No. 12½	17/64	200
8d	2½ inch...No. 11½	19/64	136
10d	3 inch...No. 10½	5/16	90
16d	3½ inch...No. 10	11/32	69
20d	4 inch...No. 9	⅜	50

CASING NAILS
Deep Countersunk Head, Diamond Point

Size	Length and Gauge	Degree of Countersunk	Dia. Head Ga.	Ap. No. to Lb.
4d	1½ inch..No. 14	32	11	489
6d	2 inch..No. 12½	32	9½	244
8d	2½ inch..No. 11½	32	8½	147
10d	3 inch..No. 10½	32	7½	96
16d	3½ inch..No. 10	32	7	73

STEEL ESCUTCHEON PINS
Oval Head, Needle Point

Carpet Tacks
No. 4 No. 6 No. 8

Upholsterer Tacks
No. 4 No. 6 No. 8

FINISHING NAILS
Brad Head, Diamond Point

Size	Length and Gauge	Dia. Head Gauge	Approx. No. to Pound
3d	1¼ inch..No. 15½	12½	880
4d	1½ inch..No. 15	12	630
6d	2 inch..No. 13	10	288
8d	2½ inch..No. 12½	9½	196
10d	3 inch..No. 11½	8½	124

Double Pointed Tacks
No. 8 No. 9 No. 10

Fig. 10. Types and sizes of nails. The letter "d" refers to weight per thousand nails.

8d FINISHING NAIL
1" — No. 20 BRAD

Rivets. Permanent fastening devices used to join parts made of sheet metal or plate steel are called rivets. Rivets are made of many different kinds of metal, such as iron, steel, copper, brass, and aluminum. They are available with flat, countersunk, button, pan and truss type heads. See Fig. 11. Specifications of rivets include diameter, length, head shape, and kind of metal.

⅛ x ¾ RD HD STEEL RIVET

Pins. Parts of an assembly may be locked into position by means of a pin.

Cotter pins are used for retaining or locking slotted nuts, ball sockets, and movable links or rods, as shown in Fig. 12.

Groove pins are straight pins made of cold drawn steel. They have a groove running the entire length of the pin. The groove is rolled or pressed into the body of the pin to provide a locking effect when the pin is driven into a drilled

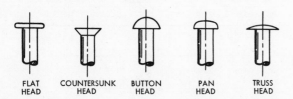

FLAT HEAD   COUNTERSUNK HEAD   BUTTON HEAD   PAN HEAD   TRUSS HEAD

Fig. 11. Rivets are used to join thin sections of materials.

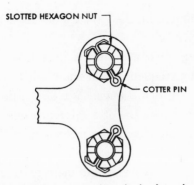

Fig.12. Cotter pins are used to lock slotted nuts, rods, or movable links.

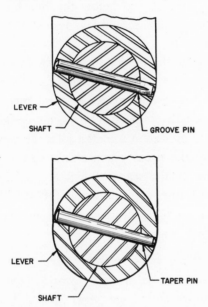

Fig. 13. Groove and taper pins have similar appearances and functions.

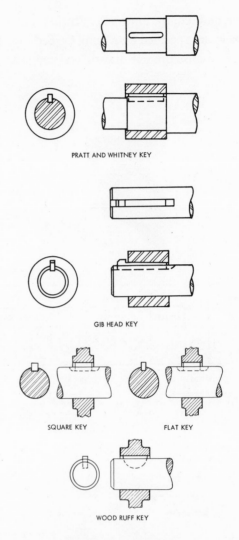

Fig. 14. Keys are used to prevent a part from rotating on its shaft.

hole. See Fig. 13. This type of pin is used for semi-permanent fastening of levers, collars, gears, and cams to shafts.

Taper pins are similar to groove pins. However, they require taper-reamed holes at assembly and depend only on taper locking which can totally disengage when minor displacement occurs. See Fig. 13.

Keys. Cylindrical parts attached to shafts, such as pulley wheels and gears, are prevented from rotating on their shafts by using a fastener called a key. Fig. 14 illustrates some of the more common types of keys.

On drawings, flat, taper, square, and gib-head keys are specified by a note giving the width,

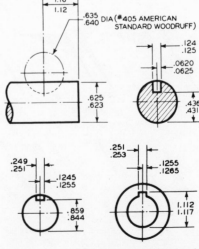

Fig. 15. Keyways are fully dimensioned; the key is seldom shown.

height, and length. Pratt and Whitney and Woodruff keys are specified by a number. See Fig. 15.

Keys are usually not drawn. The common practice is to dimension the keyways on shafts or other internal members.

Welding. When parts are joined by welding, the drawing frequently will show where and how the welds are to be made. Welding information is presented by means of a reference line with an arrowhead pointing to the welded joint. See Fig. 16.

Special symbols are used to indicate the type and size of the weld. These symbols are placed above the reference line for a far weld, or a weld on the opposite side, and below the line for a near weld. The size of the weld is indicated by a dimension number placed at the left of the weld symbol. The length of the weld is shown by a dimension at the right side of the symbol. See Fig. 16.

Soldering. Seams made on sheet metal products are usually soldered. No specific designations are used to indicate a soldered joint. However, on a drawing the usual practice is simply to run a leader line to the seam, shown graphically by a heavy black line, with the word SOLDERED. Fig. 17 illustrates some of the more common seams used in sheet metal work.

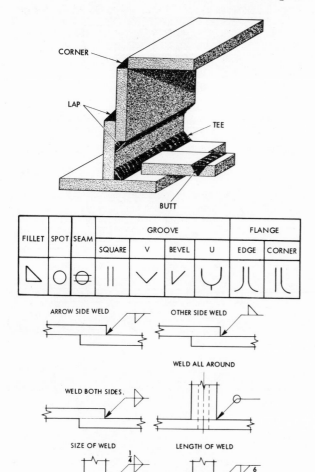

Check your knowledge of this unit by completing Self-Check for Unit 6 on pages 71 and 72.

Complete sketches of the objects on Drawing Sheets 18, 19, 20 and 21, pages 73-76.

Fig. 16. Method of designating welds on a drawing.

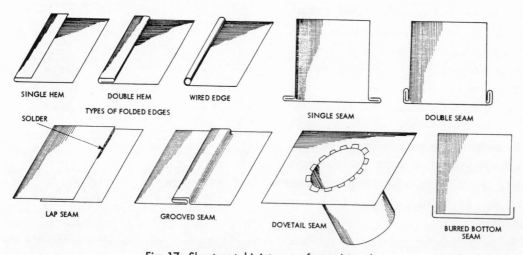

Fig. 17. Sheet metal joints are often soldered.

SELF-CHECK FOR UNIT 6

DIRECTIONS: Circle the letter T if the statement is True or the letter F if the statement is False.
Self-Check answers may be found on page 172.

1. T F Production drawings usually require detail as well as assembly drawings of the object to be made.

2. T F A production drawing is often referred to as a working drawing.

3. T F A detail drawing contains only a single view of the object.

4. T F A parts list is associated with a detail drawing.

5. T F A detail drawing and an assembly drawing may sometimes be shown on one sheet.

6. T F UNC threads are used on bolts and screws for general application.

7. T F A class 3 thread possesses the largest allowance between the mating screw and nut.

8. T F When a thread class is marked 2B it represents an external thread.

9. T F The easiest way to show threads on a drawing is by the simplified representation method.

10. T F A nut is used with a stud.

11. T F The specifications for this bolt are correctly shown:
$1/4 — 2A \times 1\,1/4$ HEX BOLT

12. T F A machine screw is the same as a cap screw except that it is usually smaller in size.

13. T F A wood screw marked No. 8 means that the screw has a diameter of a No. 8 wire.

14. T F The term penny (d) is used to designate the size of self-tapping screws.

15. T F Specification of rivets must show the diameter, length, head shape and material.

16. T F When a weld symbol is used to designate a welded joint, the number shown to the left of the weld symbol indicates the size of the weld.

17. T F The three types of pins you have studied can be used to lock two pieces together so that one will not move with respect to the other.

18. T F The most dependable type of pin is the taper pin.

19. T F On assembly drawings keys must be fully dimensioned.

20. T F On drawings where keys are dimensioned, the dimensions are either specified by a note or are specified by a number.

21. T F The usual practice is to dimension keyways and NOT the key itself.

22. T F Keys are always shown on a drawing.

23. T F A parts list gives the type and number of pieces required.

24. T F Thread class refers to the degree of coarseness of the thread.

25. T F There are many different types of assembly drawings in addition to the general assembly drawing.

Score_____

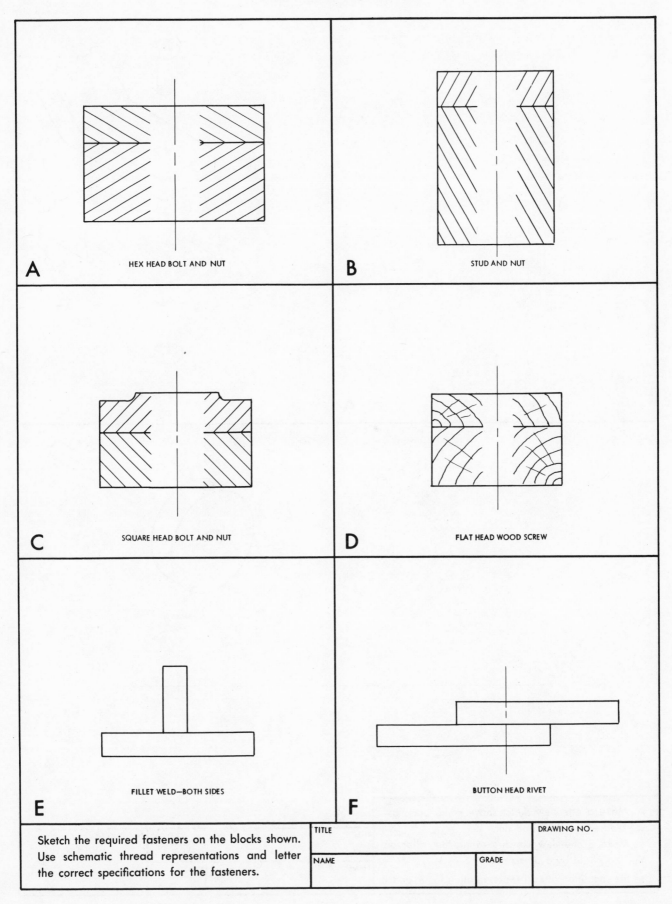

A　HEX HEAD BOLT AND NUT

B　STUD AND NUT

C　SQUARE HEAD BOLT AND NUT

D　FLAT HEAD WOOD SCREW

E　FILLET WELD—BOTH SIDES

F　BUTTON HEAD RIVET

Sketch the required fasteners on the blocks shown. Use schematic thread representations and letter the correct specifications for the fasteners.

TITLE		DRAWING NO.
NAME	GRADE	

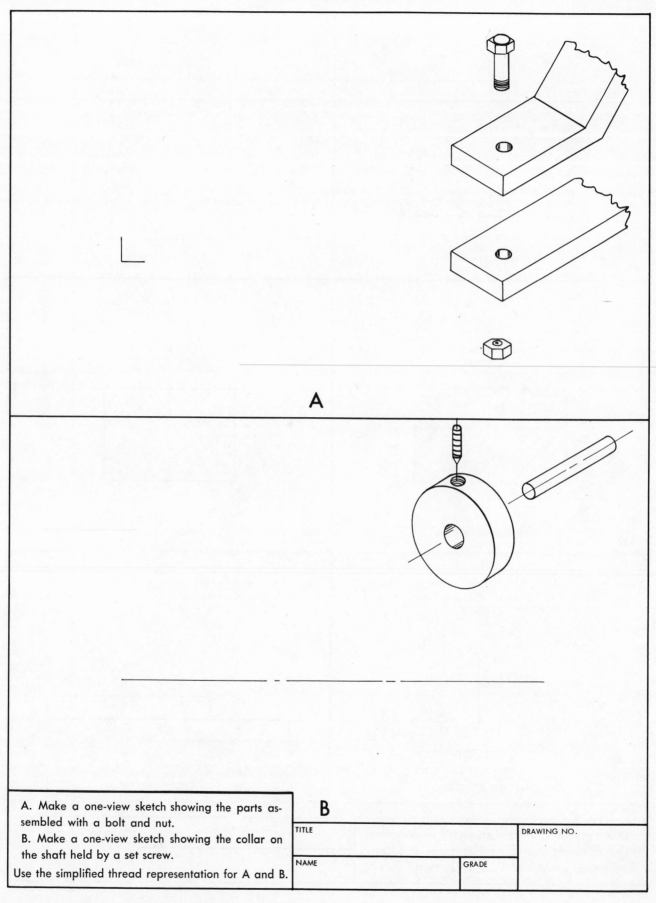

A

B

A. Make a one-view sketch showing the parts assembled with a bolt and nut.
B. Make a one-view sketch showing the collar on
the shaft held by a set screw.
Use the simplified thread representation for A and B.

TITLE		DRAWING NO.
NAME	GRADE	

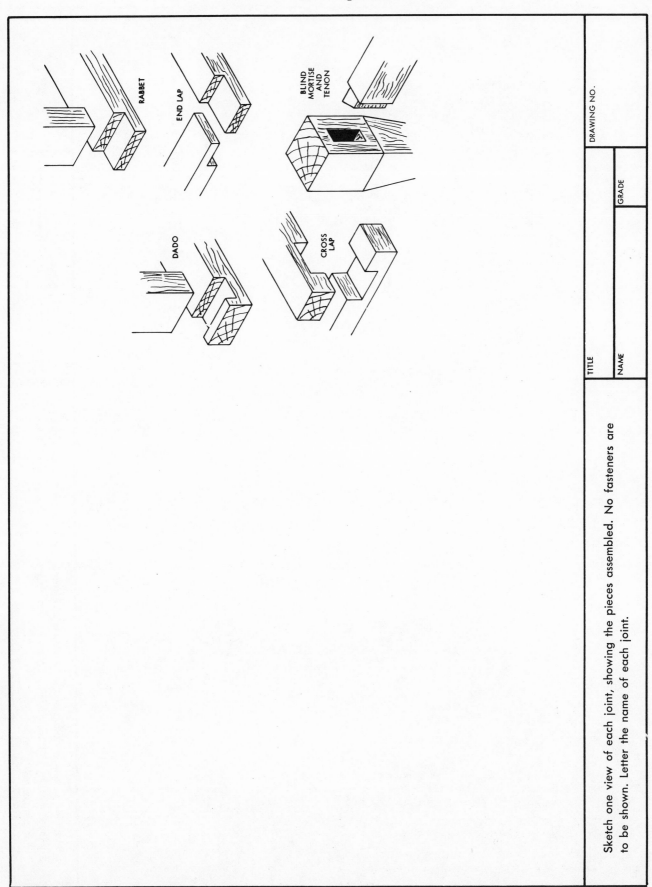

RABBET

END LAP

BLIND MORTISE AND TENON

DADO

CROSS LAP

DRAWING NO.

GRADE

TITLE

NAME

Sketch one view of each joint, showing the pieces assembled. No fasteners are to be shown. Letter the name of each joint.

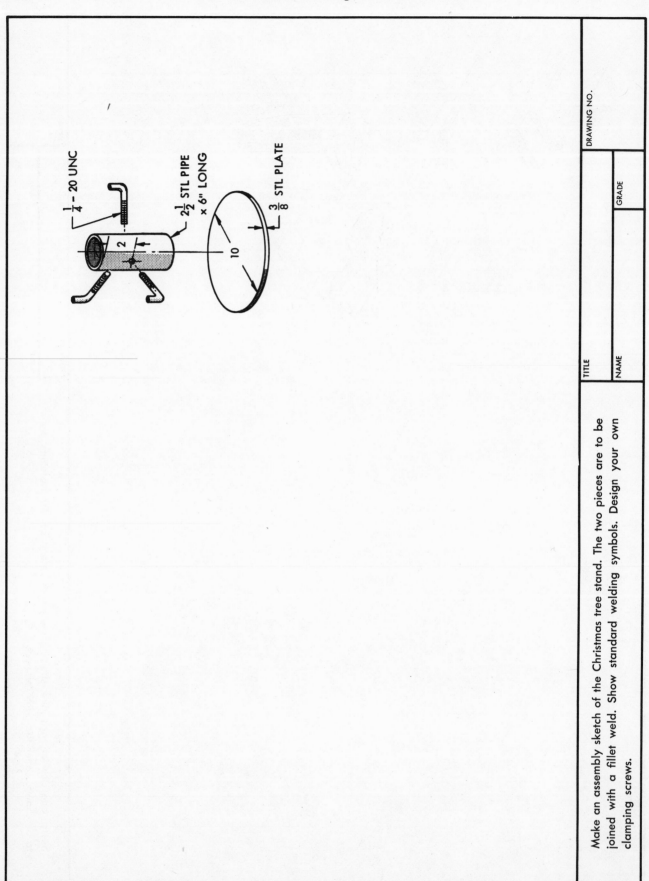

$\frac{1}{4}$ - 20 UNC

$2\frac{1}{2}$ STL PIPE
x 6" LONG

2

$\frac{3}{8}$ STL PLATE

10

DRAWING NO.

GRADE

TITLE

NAME

Make an assembly sketch of the Christmas tree stand. The two pieces are to be joined with a fillet weld. Show standard welding symbols. Design your own clamping screws.

Sketching Pictorial Drawings

Pictorial drawings are life-like drawings of objects, drawn as the viewer might expect to see the actual objects. The line of sight is never directly in front of any one side, so that more than one side may be seen. Usually, a pictorial drawing shows two sides and the top of the object.

Pictorial drawings are becoming more and more widely used by newspapers, magazines, and catalogs. The viewer receives a much clearer idea of what the object looks like, since the drawing is largely self-explanatory. The three common types of pictorial drawings are the isometric, oblique, and the perspective.

Isometric drawings

An isometric drawing clearly shows three surfaces of an object. See Fig. 1. The drawing consists of three axes, one of which is vertical and the other two drawn to the right and left at an angle of 30° to the horizontal. See Fig. 2.

The object can be rotated so that either the right or left side is visible, as in Fig. 3. Whether the object is drawn with its main surfaces extending to

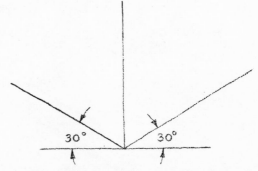

Fig. 2. Two of the axes of an isometric drawing are at 30° to the horizontal.

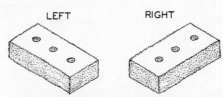

Fig. 3. In an isometric drawing the object can be rotated about the vertical axis to the left or right.

the right or left depends entirely upon which side is most desirable to show. Hidden lines are usually not included in pictorial drawings.

To make an isometric drawing, proceed as follows:

1. Draw a vertical line. From the top or base of this line extend two slanted lines at an angle of 30° to the horizontal. See A in Fig. 4.
2. Lay out the actual width, length, and height on these three lines. See B in Fig. 4. Complete each surface by drawing the necessary lines parallel to the axes, as shown in C and D of Fig. 4.

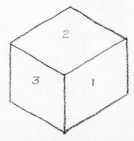

Fig. 1. An isometric drawing clearly shows three surfaces.

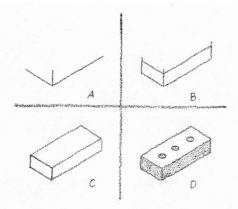

Fig. 4. Steps in preparing an isometric drawing.

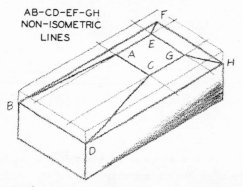

Fig. 5. To draw non-isometric lines, locate the endpoints of the lines on isometric surfaces.

3. When an object has slanted lines that do not run parallel to the axes, their true lengths cannot be measured like regular isometric lines. Therefore, it is necessary to locate the extreme ends of these lines first. Thus in Fig. 5, points A, B, C, and D are located and then lines AB and CD are drawn.

4. If the object has an irregular curve as in Fig. 6, the curve is constructed by drawing a series of isometric lines. The points are plotted by measuring along these isometric lines. Finally, the curve is drawn through these points.

5. To draw an isometric circle or arc, construct an isometric square of the required size. Mark the center of each side of the square. See Fig. 7 (Left). From centers A and B draw lines to the corner Y. From opposite sides C and D draw lines to the opposite corner W. See Fig. 7 (Center).

Using r as the radius and the points AY-DW and BY-CW as centers, draw arcs AD and BC. Using R as the radius and the corners W and Y as centers, draw arcs DC and AB. See Fig. 7 (Right). Use the same pro-

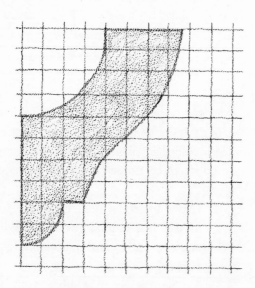

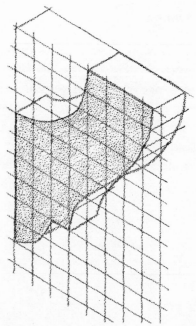

Fig. 6. Drawing an irregular curve in isometric is accomplished by constructing an isometric grid and projecting points onto this grid.

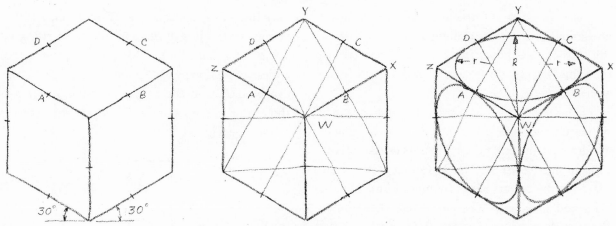

Fig. 7. Drawing an isometric circle with the four-center system.

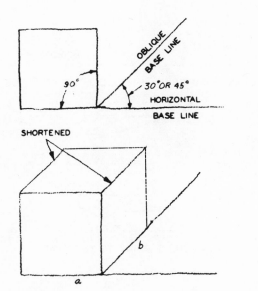

Fig. 8. An oblique drawing has one face parallel to the plane of projection. This face shows true size and shape, the other faces are foreshortened.

cedure to draw circles on the other faces of the isometric object. This procedure is called the *four-center system*.

Oblique drawings

The oblique drawing, like the isometric drawing, has three axes. Two of these axes are at 90° to each other, and the third is generally at 30° or 45° to the horizontal. See Fig. 8 (Top).

The oblique drawing resembles the isometric except that one face of the object is parallel to the plane of projection. Only that face appears in true size and shape. The remaining two faces are drawn short to avoid the appearance of distortion. See Fig. 8 (Bottom). Notice that side *b* is actually drawn shorter than side *a*, although the object represents a cube, and the sides are really equal.

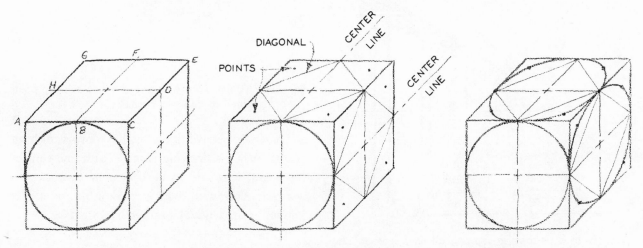

Fig. 9. An alternate method for constructing an oblique circle is the use of triangles.

Procedures for constructing oblique drawings are similar to those for making an isometric drawing. Although oblique circles can be constructed in the same way as are isometric circles, there is an alternate method.

The oblique square is constructed and the circle drawn on the face. See Fig. 9 (Left). Diagonals are drawn connecting the midpoints of the sides in top and side views. See Fig. 9 (Center). Points are placed in the centers of the triangles formed by the sides and diagonals. The circles are then drawn through these points. See Fig. 9 (Right).

Perspective drawings

The perspective drawing is the most realistic of the pictorial drawings. The perspective drawing is based on the observation that objects seem smaller the farther away they are from the viewer. Street lights and telephone poles seem to be decreasing until they disappear at the horizon. Parallel lines all seem to merge at a point on the horizon.

Notice in Fig. 10 how the railroad tracks and telephone poles all seem to merge and disappear at a point on the horizon.

The point where all lines appear to meet is known as the *vanishing point*. This vanishing point is located on the horizon, which is an imagi-

nary line in the distance and at eye level. All lines below eye level have the appearance of rising upward to the horizon, and all lines above eye level appear to go downward to the horizon. The object can be drawn so the vanishing point is either to the left or to the right, or directly in the center of vision. See Fig. 11.

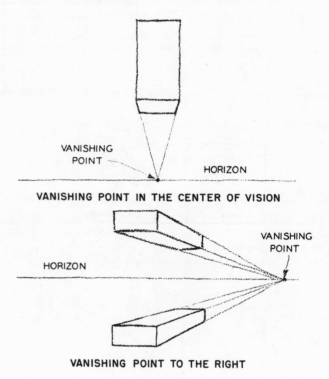

VANISHING POINT IN THE CENTER OF VISION

VANISHING POINT TO THE RIGHT

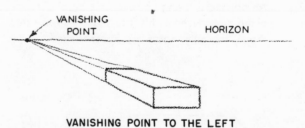

VANISHING POINT TO THE LEFT

Fig. 11. An object can be drawn with the vanishing point located in different positions.

Some perspective drawings have two vanishing points. When a drawing is made with one vanishing point, it is said to be drawn in *parallel perspective*. A drawing with two vanishing points is known as an *angular perspective*. See Fig. 12.

To make a simple perspective drawing, proceed as follows:

Fig. 10. In a perspective drawing all lines seem to disappear at the horizon.

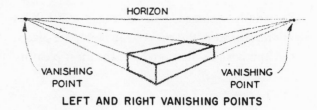

Fig. 12. A perspective may have two vanishing points.

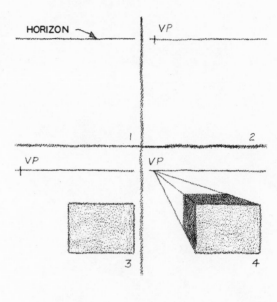

Fig. 13. One vanishing point is used in drawing a one-point (parallel) perspective.

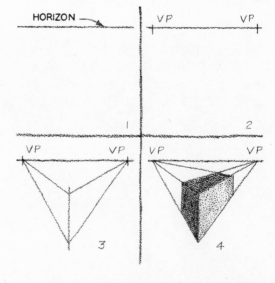

Fig. 14. Two vanishing points are used in drawing a two-point (angular) perspective.

1. Assume the location of the horizon line.
2. Locate the position of the vanishing points.
3. For a parallel perspective, draw a front view of the object. This will be a true view. See Fig. 13. If an angular perspective is to be made, draw a vertical line and on it lay off the full or scaled height of the object. See Fig. 14.

4. From the front view or vertical line, draw light lines back toward the vanishing points.
5. The points designating the length or depth of the object may be found by a projection method. However, since this is a more involved procedure, it is used only when an accurate mechanical perspective is made. For most purposes, location points for depth are

simply assumed, that is, the vertical lines representing the ends of the object are placed in a position that produces the most pleasing effect.

Check your knowledge of this unit by completing the Self-Check for Unit 7 on pages 83 and 84.

Make sketches of the objects shown on Drawing Sheets 22-29, pages 85-92. Follow the instructions given at the bottom of each drawing sheet.

SELF-CHECK FOR UNIT 7

DIRECTIONS: Circle the letter which corresponds to the best answer to each of the following questions. *Self-Check answers may be found on page 172.*

1. In an isometric drawing the width, height, and depth of the object are

   a. measured only along the 30° lines of projection
   b. measured on the three projecting axes
   c. judged by eye to achieve realistic proportions
   d. not particularly important for a balanced effect

2. Nonisometric lines are lines which

   a. can be drawn parallel to regular isometric axes
   b. can be measured in the same manner as normal isometric lines
   c. cannot be measured in the same manner as normal isometric lines
   d. can be shown in their true length in their respective views

3. Isometric circles are

   a. drawn in their true shape from a center point
   b. drawn by the four-center system
   c. laid off with an irregular curve
   d. plotted by transferring them from an orthographic view

4. In an isometric drawing an object can be rotated so that

   a. only one surface, the left side, is visible
   b. only one surface, the right side, is visible
   c. both the right and left sides are visible at the same time
   d. either the right or left side is stressed

5. In an oblique drawing, the number of surfaces that appear in true size and shape are

   a. one
   b. three
   c. two
   d. all surfaces

6. A circle in the front surface of an oblique drawing appears in

   a. elliptical form
   b. true shape
   c. distorted form
   d. conical form

7. In a perspective drawing, horizontal lines parallel to the horizon are sketched

   a. parallel to the horizon
   b. at a 45° angle
   c. at any angle to the horizon
   d. parallel to the receding lines

8. Vanishing points in perspective drawings are always

   a. below the horizon
   b. above the horizon
   c. on the horizon
   d. above and below the horizon

84

9. A parallel perspective drawing has

   a. two vanishing points
   b. a vanishing point above and below the horizon
   c. one vanishing point
   d. a vanishing point extending to the right and left

10. For relatively simple perspective drawings the depth of the object

   a. is measured on receding lines
   b. can be assumed
   c. is disregarded completely
   d. can be assumed to be the same as the front face

Score_____

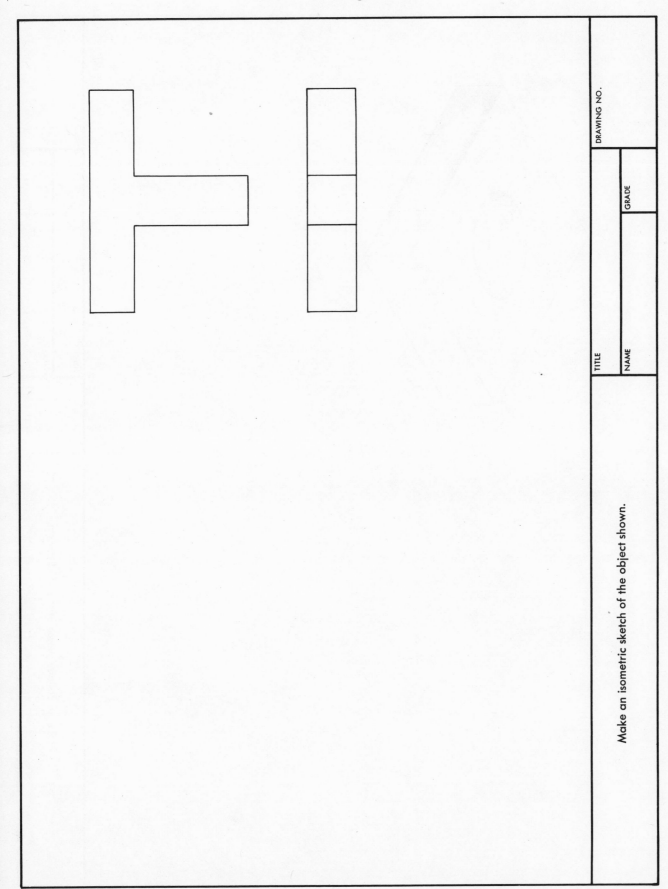

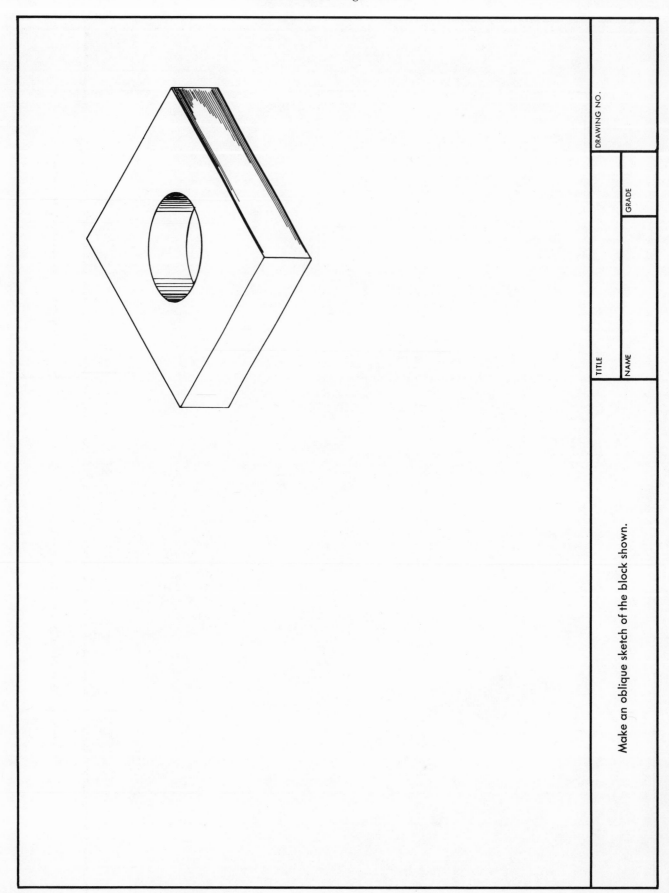

DRAWING NO.

GRADE

TITLE

NAME

Make an oblique sketch of the block shown.

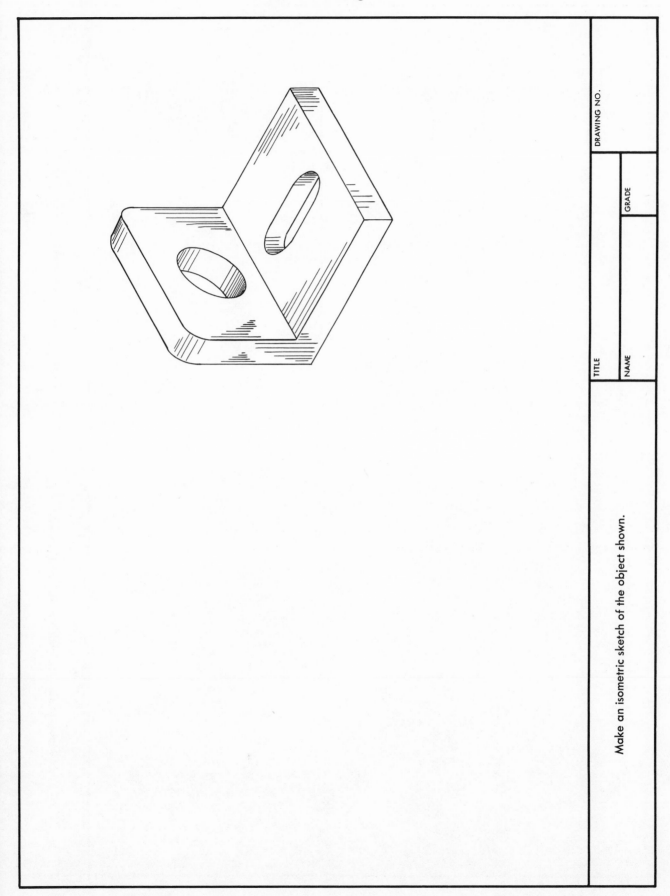

DRAWING NO.

GRADE

TITLE

NAME

Make an isometric sketch of the object shown.

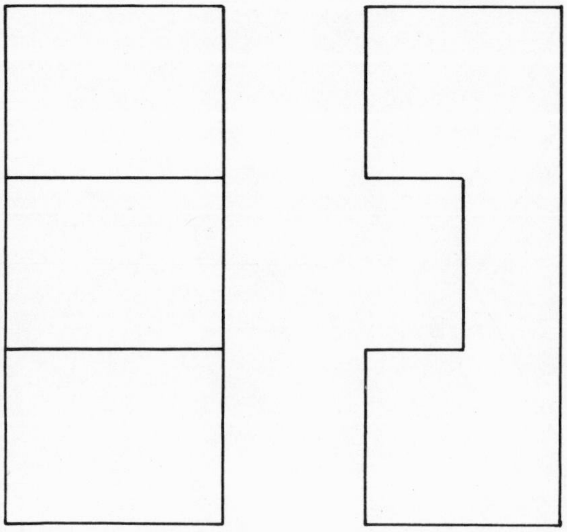

Make an isometric sketch of the block shown.

TITLE

DRAWING NO.

NAME

GRADE

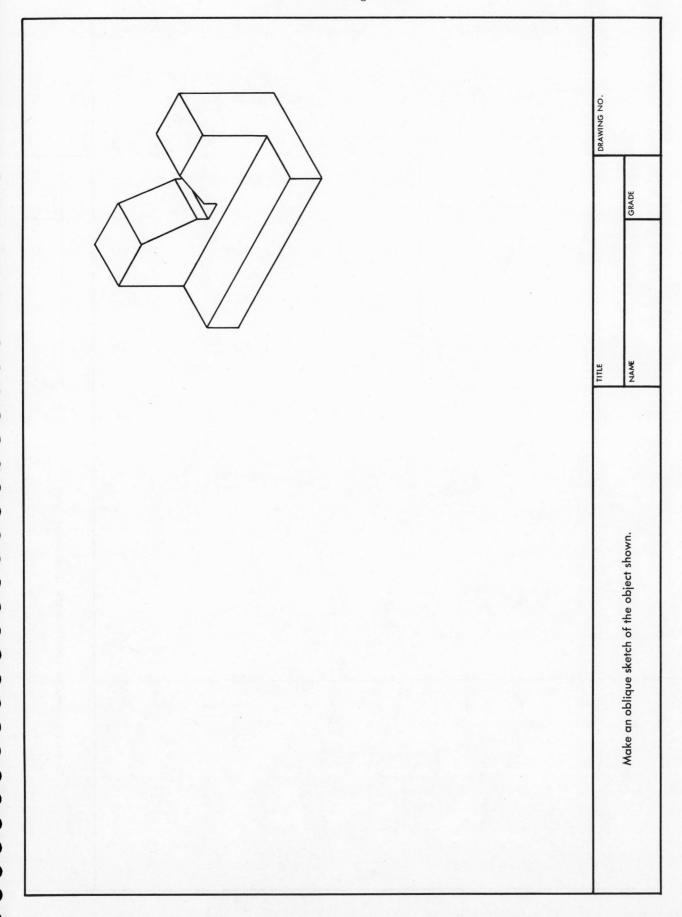

GRADE

TITLE

NAME

Make an oblique sketch of the object shown.

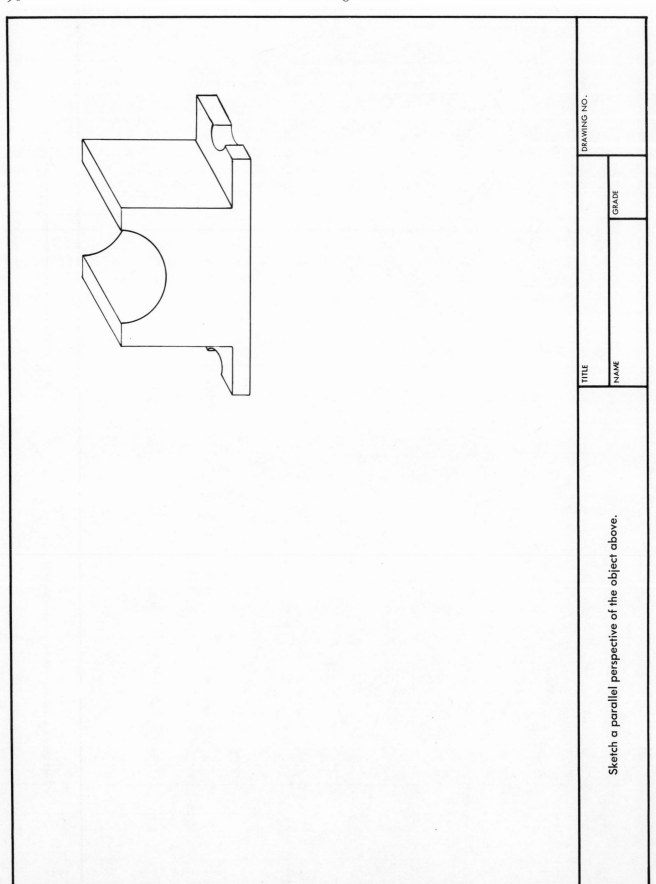

DRAWING NO.

GRADE

TITLE

NAME

Sketch a parallel perspective of the object above.

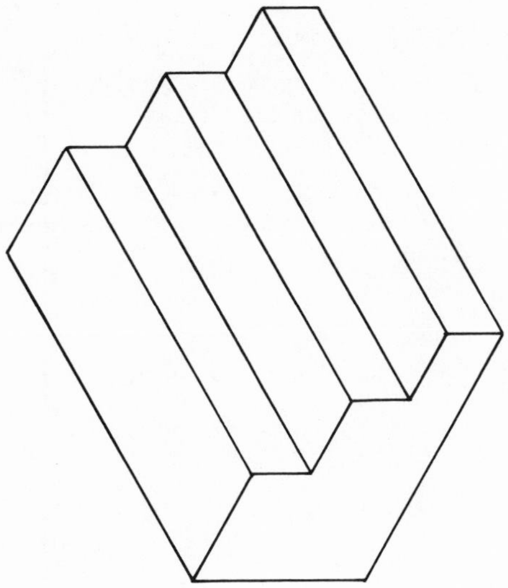

DRAWING NO.

GRADE

TITLE

NAME

Sketch an angular perspective of the flight of stairs shown.

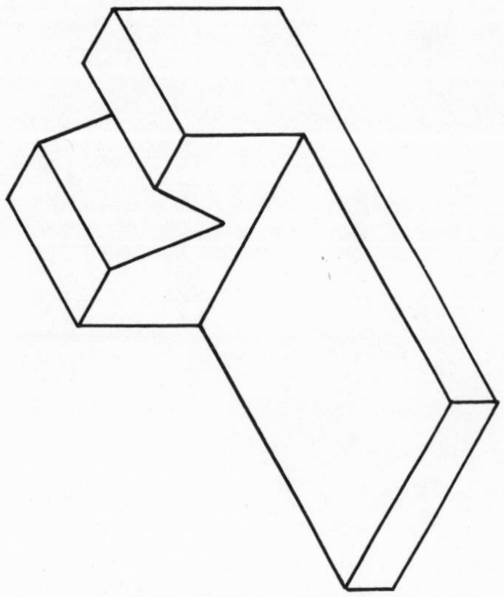

DRAWING NO.

TITLE

NAME

GRADE

Sketch a parallel perspective of the object shown.

Making
Instrument Drawings

Your study of drawings so far has dealt entirely with making freehand sketches. In this unit you will have an opportunity to use some of the basic drafting instruments in preparing working and pictorial drawings.

Drawing board

An 18″ x 24″ drawing board is generally used by most beginners. This board is made of white pine or some other soft wood.

The drawing paper is fastened to the board with drafting tape. First a piece of tape is placed across the upper left hand corner of the sheet. A T-square is then moved up to align the bottom edge or printed border line of the paper and another strip of tape fastened on the upper right hand corner. This is followed by taping the two bottom corners. See Fig. 1.

Pencil

As you learned in Unit 2, drawing pencils are available with many kinds of lead. Draftsmen usually prefer a medium lead for instrument drawing because erasing is easier whenever corrections must be made. A hard lead leaves deep impressions or grooves which cannot be removed. For most of your instrument drawings, you should use an H or 2H pencil.

Always keep your pencil correctly pointed to produce clear sharp lines. After sharpening the pencil in the pencil sharpener, it is a good idea to roll the point on a sandpaper pad. See Fig. 2.

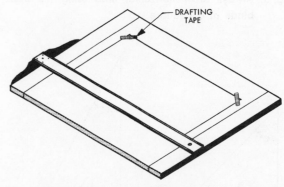

Fig. 1. Begin fastening the drawing sheet to the board by taping the top of the sheet first.

Fig. 2. Form the pencil point on a sandpaper board.

T-Square

The T-square consists of a long straight strip, called the blade, mounted at right angles on a short piece, called the head. The T-square is used to draw horizontal lines and to serve as a base for the triangles when vertical and slanted lines are drawn. The T-square is moved up or down to the required position by sliding the head along the edge of the board with your left hand. See Fig. 3. Drawings are kept clean if the T-square is lifted slightly when moved to keep it from rubbing across drawn lines.

Fig. 3. Always draw horizontal lines using the upper blade of a T-square.

Fig. 5. When drawing vertical lines with the aid of a triangle, run the pencil up the edge of the triangle.

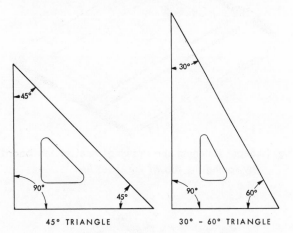

45° TRIANGLE   30° - 60° TRIANGLE

Fig. 4. The standard triangles are the 30°-60° angle and the 45° angle.

Horizontal lines are drawn along the upper edge of the blade with the pencil moving from left to right. As the line is drawn, tilt the pencil in the direction the line is made and slightly away from the edge of the T-square. Use a uniform pressure on the pencil and rotate it slightly.

Triangles

Right-angle triangles are used in combination with the T-square to draw vertical and slanted lines. The basic triangles are the 45° and the 30°-60° as shown in Fig. 4.

To draw vertical lines, rest the triangle against the T-square. Slide the triangle to the required position and draw the line by starting at the bottom and running the pencil upward while holding it

slightly away from the edge of the triangle. See Fig. 5.

To draw slanted lines, place the triangle with the angle required against the edge of the T-square. With the standard triangles you can draw lines at angles of 30°, 45°, 60°, and 90°. Other angles can be drawn by combining the two triangles as in Fig. 6. Notice in Fig. 7 the direction the pencil should be moved in drawing various inclined lines.

Protractor

Sometimes you will have to draw slanted lines at an angle which cannot be obtained with triangles. In such cases you will have to use a protractor. A protractor has a semicircular shape divided into 180 units, or degrees. Two scales are shown on the protractor with each scale having units running from 0 to 180 degrees. The outer scale is used to draw angular lines that extend to the left and the inner scale for lines that must be drawn to the right. See Fig. 8. This is how a protractor is used:

1. Place the protractor on the T-square with its center point on the mark from where the line is to be drawn.
2. Find the desired angle on the protractor, using either the inner or outer scale, and mark it with a point.
3. Remove the protractor and, with the edge of a triangle, draw a line connecting the two points. See Fig. 8.

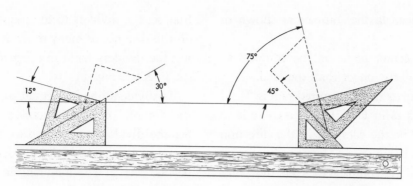

Fig. 6. Types of inclined lines that can be drawn with triangles.

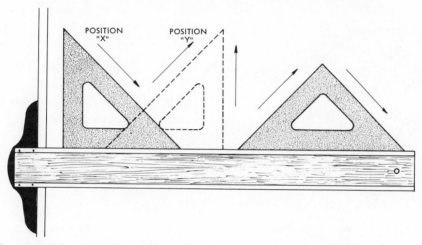

Fig. 7. The direction of the pencil in drawing inclined lines is such that the pencil is pulled and never pushed.

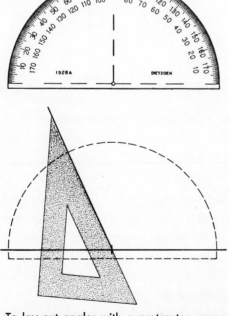

Fig. 8. To lay out angles with a protractor, measure the angle desired, mark it, and draw a line from the mark to the origin.

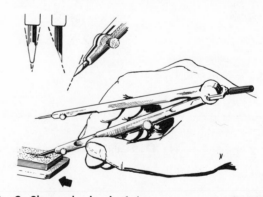

Fig. 9. Shape the lead of the compass to a fine bevel point.

Compass

A compass is used to draw circles or arcs. The large compass is usually better to draw circles whose radii range from 1 to 6 inches. For very small circles or arcs, the bow compass is more convenient.

To draw circles or arcs proceed as follows:

1. Shape the lead in the compass as shown in Fig. 9.
2. Adjust the center point on the compass so that it is slightly longer than the lead.
3. Set the compass to the required radius. Hold the compass as in Fig. 10 and revolve it to the right. Tilt the compass in the direction the circle is being drawn. Exert only enough pressure on the compass to keep the point from coming out of the paper.

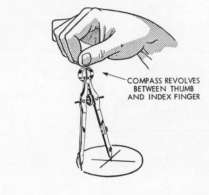

COMPASS REVOLVES BETWEEN THUMB AND INDEX FINGER

line, set the dividers to the required distance. Step off this distance as many times as desired by swinging the dividers from one leg to the other along the line. See Fig. 11.

To divide a line into equal parts, first determine the number of equal distances that are required. Set the dividers to the approximate distance and space off the parts. If the dividers fall short of or beyond the given line, lengthen or shorten the dividers and repeat the operation.

Fig. 10. When drawing a circle with a compass, turn the compass clockwise.

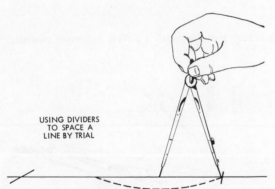

USING DIVIDERS TO SPACE A LINE BY TRIAL

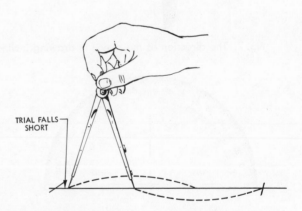

TRIAL FALLS SHORT

Fig. 11. Slightly prick the paper when spacing off equal distances with dividers.

Dividers

Dividers are similar to compasses, except that both legs have needle points. They are used primarily for transferring measurements and dividing lines into equal parts.

To lay off distances, set the dividers to the correct length, and then transfer the measurements to the drawing by pricking the drawing surface very lightly with the points of the dividers.

To measure off a series of equal distances on a

Curves

When you must draw curved lines which are not arcs of circles, you will need a special guide such as an irregular curve. To use an irregular curve, follow this procedure:

1. Lay out a series of points to indicate the shape of the curve.
2. Locate a part of the irregular curve that fits as many points as possible and scribe a line. See Fig. 12.

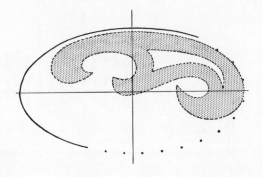

Fig. 12. When drawing an irregular curve, line up as many points as possible at one time with the curve.

3. Move the curve to the next position and draw a line to connect with the first scribed line. Continue this procedure until you have completed the curved line. Be sure the places where the curved sections are joined are uniform with the rest of the curved line.

Measuring scales

Several types of measuring scales are used by draftsmen, such as the architect's, the mechanical engineer's and the civil engineer's scales. For this course you will be using the architect's scale. See Fig. 13. Although its name would indicate that this scale is intended primarily for architectural drawing, actually it may be used for making many types of drawings.

If you examine the architect's scale, you will find that each edge has a different measuring unit. These units are designed so the part being drawn can be represented on standard-size paper. Some objects are so large that it is practically impossible to draw them full size while other objects are often too small to draw them to actual size. By using the various measuring units on the architect's scale, and drawing the object to scale, it is possible to draw objects to any size so they will fit on the drawing sheet.

You will find that one edge of the architect's scale is divided into inches with subdivisions of sixteenths. See Fig. 14. The other edges have units which represent different proportions in terms of feet and inches. These measuring units are labeled 3, 1-1/2, 1, 3/4, 1/2, 3/8, 1/4, 3/16, 1/8, and

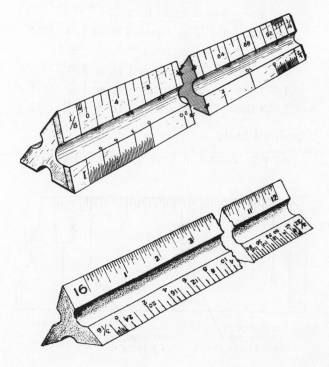

Fig. 13. The architect's scale has several measuring edges so that objects can be drawn to different sizes, or scales.

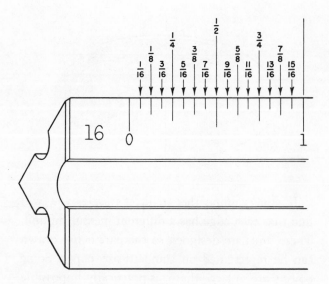

Fig. 14. Divisions of the one inch unit on the architect's scale.

3/32. The edge marked with a 3 contains a scale on which 3 inches is equal to one foot. The edge having 1-1/2 contains the scale on which 1-1/2 inches equals one foot, and so on.

When the various measuring units are used, a drawing is said to be drawn to scale, such as full size, quarter-size, etc. The scales used most often for full size and reduced drawings are:

Full size—12 in. equals 1 ft. or $1'' = 1''$
3/4 size—9 in. equals 1 ft. or $3/4'' = 1''$
1/2 size—6 in. equals 1 ft. or $1/2'' = 1''$
1/4 size—3 in. equals 1 ft. or $1/4'' = 1''$
1/8 size—1-1/2 in. equals 1 ft. or $1/8'' = 1''$
1/12 size—1 in. equals 1 ft. or $1'' = 1'$
1/16 size—3/4 in. equals 1 ft. or $3/4'' = 1'$
1/24 size—1/2 in. equals 1 ft. or $1/2'' = 1'$
1/48 size—1/4 in. equals 1 ft. or $1/4'' = 1'$

Decimal Scale

As you learned in Unit 5, most industries are now using decimals rather fractional sizes in dimensioning drawings. The scale used for this purpose is graduated in decimal units.

Decimal scales are made with decimal units divided into 10, 20, 30, 40, 50, and 60 parts to the inch. Thus the edge marked 10 simply means that one inch has been subdivided into 10 equal parts. A scale where the inch is divided into 50 parts, which is the one most often used in industry, means each part is 1/50 of an inch, or a decimal equivalent of .02 inch. Notice in Figs. 15 and 16 how this scale can be used to produce full, one-half and one-quarter size drawings.

Drawing an object to scale

Let us assume that you want to draw an object which has to be reduced in size. Suppose, for example, that your drawing paper is such that you can lay out the object to a scale of $3/4'' = 1'-0''$.

Fig. 15. How to use the decimal scale for a full-size drawing.

5

0   1

THIS INDICATES
5/10 OR .500   THIS INDICATES   THESE LINES
OR 1/2 INCH   A FULL INCH   SHOW 1/10
                                             OR .100 INCH

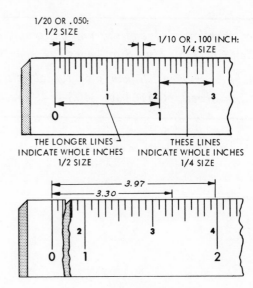

Fig. 16. How to use a decimal scale for half-size and one-quarter size drawings.

Fig. 17. Notice in this use of the three-quarter scale that the number of whole units is measured from the 0 point.

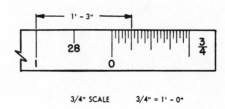

Fig. 18. Notice in this use of the one-quarter scale that fractional parts of a unit are measured from the right of the 0 point.

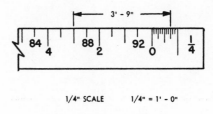

Find the scale that is marked ¾. Remember that each unit line on this scale represents one foot.

Observe, too, that on the end of the scale the units are divided into smaller divisions. These smaller divisions represent inches and fractions of an inch. See Fig. 17. The scale on the end simply takes a large unit and subdivides it into smaller units to make it possible to lay off distances that are in fractional parts of a foot such as 1'-3", 1'-9", etc.

To use another example: assume that you want to make a drawing ¼ size. First determine the largest unit that is to be laid off, such as 3'-9". See

Fig. 18. The unit 3' then, is the largest foot dimension. Locate the scale marked 1/4 and on it find the 3' division mark. Then, reading back from the 0 into the smaller unit scale on the end, find the 9" mark. From this point to the 3' division will be our dimension to be marked off on the paper.

If a drawing of an object has to be enlarged several times, you might select the scale marked with a 3. In this case the scale is used so that each division represents one inch. This produces a drawing that is three times the actual size of the object. The 1½ scale may be used for the same purpose.

Check your knowledge of this unit by completing the Self-Check for Unit 8 on pages 101 and 102.

Make instrument drawings of the objects on Drawing Sheets 30-41, pages 103-114. Practice using the instruments on scrap paper before beginning. Follow the instructions given at the bottom of each sheet.

SELF-CHECK FOR UNIT 8

PART 1

DIRECTIONS: Circle the letter T if the statement is True or the letter F if the statement is False.
Self-Check answers may be found on page 172.

1. T F Only two corners of the drawing paper should be taped to the board.

2. T F A hard lead pencil is best for instrument drawing.

3. T F In using a T-square, the pencil should move from left to right in drawing horizontal lines.

4. T F When using triangles, vertical lines are drawn by starting at the top and moving downward.

5. T F By combining the 30°-60° and 45° triangles it is possible to lay out a 15° angle.

6. T F On a protractor, the outer scale is used to lay out angles that extend to the left.

7. T F The lead on the compass should be conical in shape.

8. T F The lead on the compass leg should be slightly longer than the center point.

9. T F In drawing an irregular curve, a series of points are first located and then a curve is used to connect these points.

10. T F The various measuring edges on the architect's scale permit drawing objects to different sizes.

11. T F The 3/4 scale may be used to represent 3/4″ = 1″ or 3/4″ = 12″

12. T F The small divisions on the end of the scale are used to find the fractional part of a unit.

PART 2

DIRECTIONS: Circle the letter which corresponds to the best answer to each of the following questions. *Self-Check answers may be found on page 172.*

1. If a 3 inch line on a drawing represents an edge of an object 3 feet long, the scale being used is
   a. 1/10
   b. 1/3
   c. 1/100
   d. 1/12

2. When drawing an object 1/4 scale, a side of the object which is 10 inches long will be drawn:
   a. 40 inches long
   b. 4 inches long
   c. 2½ inches long
   d. 5 inches long

3. An object 12 feet long, when drawn in 3/4 scale, will be shown on a drawing as a line
   a. 12 inches long
   b. 8 inches long
   c. 9 inches long
   d. 6 inches long

4. When using the edge of the architect's scale marked 3, an object which is 3 inches long is shown on a drawing as a line
   a. 1 inch long
   b. 3 inches long
   c. 10 inches long
   d. 9 inches long

5. Which of the following scales does not mean a reduction in size?
   a. 1-1/2
   b. 1/4
   c. 2/3
   d. 1/2

6. An object which is dimensioned as 14 inches long in half scale is really
   a. 28 inches
   b. 7-1/2 inches
   c. 1 inch
   d. 14 inches

7. If three views of an object 20 inches wide, 12 inches high, and 18 inches deep are to be shown on a drawing sheet which is 8½ x 11, the object will be best drawn in
   a. full scale
   b. half scale
   c. double scale
   d. one-eighth scale

8. An object 4 x 4 x 4, when drawn in 1½ scale will appear on the drawing as an object
   a. 3 x 3 x 3
   b. 4 x 4 x 4
   c. 6 x 6 x 6
   d. 9 x 9 x 9

9. The decimal unit most commonly used in industry is the one in which the inch is divided into
   a. 40 parts
   b. 10 parts
   c. 30 parts
   d. 50 parts

10. In decimal scale where the inch is divided into 50 parts, each unit represents what part of an inch?
    a. .02
    b. .50
    c. .100
    d. 1/10

Score_____

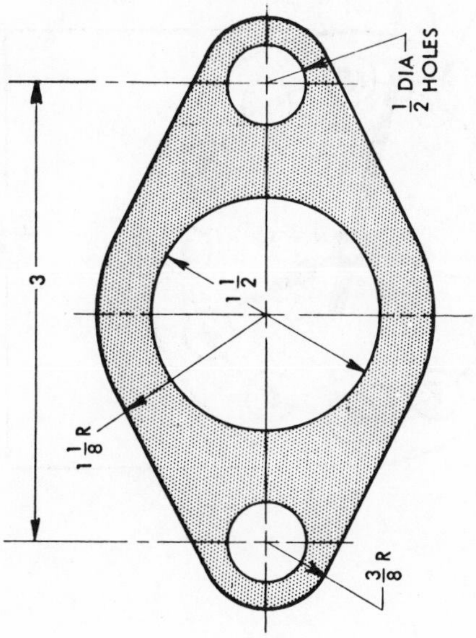

$\frac{1}{2}$ DIA 2 HOLES

3

$1\frac{1}{2}$

$1\frac{1}{8}$R

$\frac{3}{8}$R

DRAWING NO.

GRADE

TITLE

NAME

Make a one-view instrument drawing and dimension completely. Scale: full size.

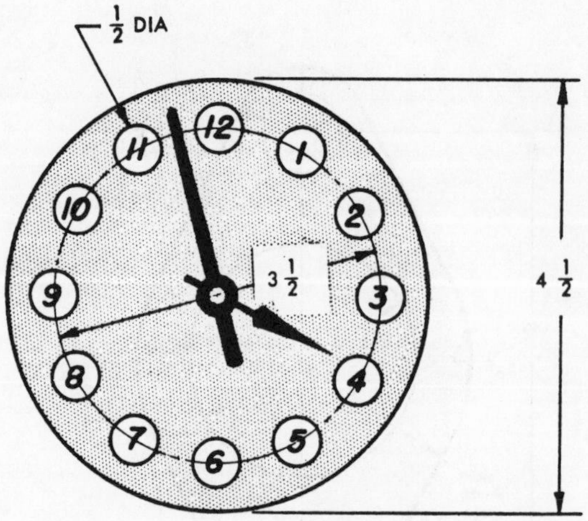

$\frac{1}{2}$ DIA

$3\frac{1}{2}$

$4\frac{1}{2}$

Make a one-view instrument drawing of the clock face. Include all required dimensions. Scale: full size.

TITLE		DRAWING NO.
NAME	GRADE	

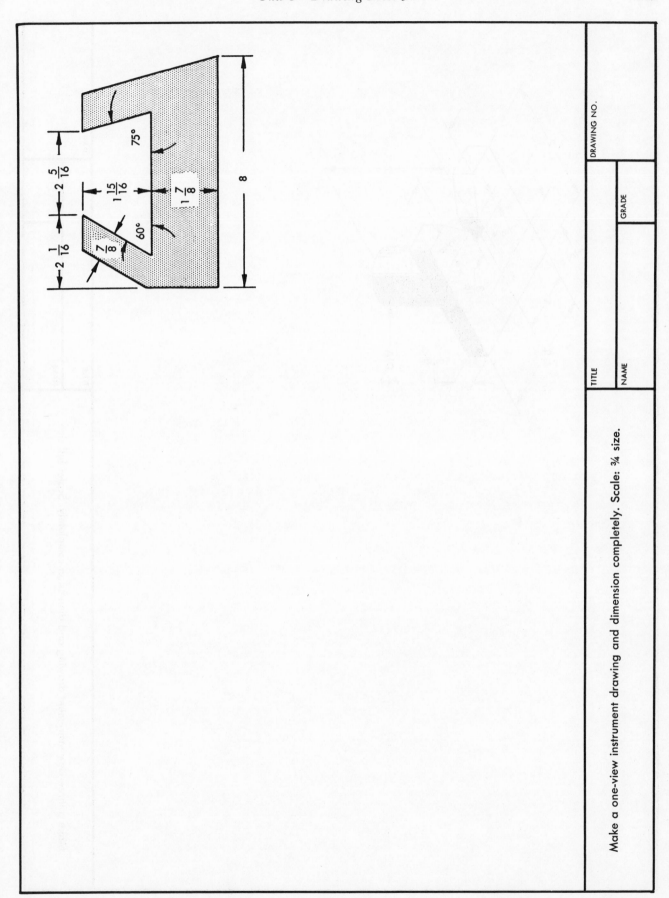

Make a one-view instrument drawing and dimension completely. Scale: ¾ size.

TITLE

DRAWING NO.

NAME

GRADE

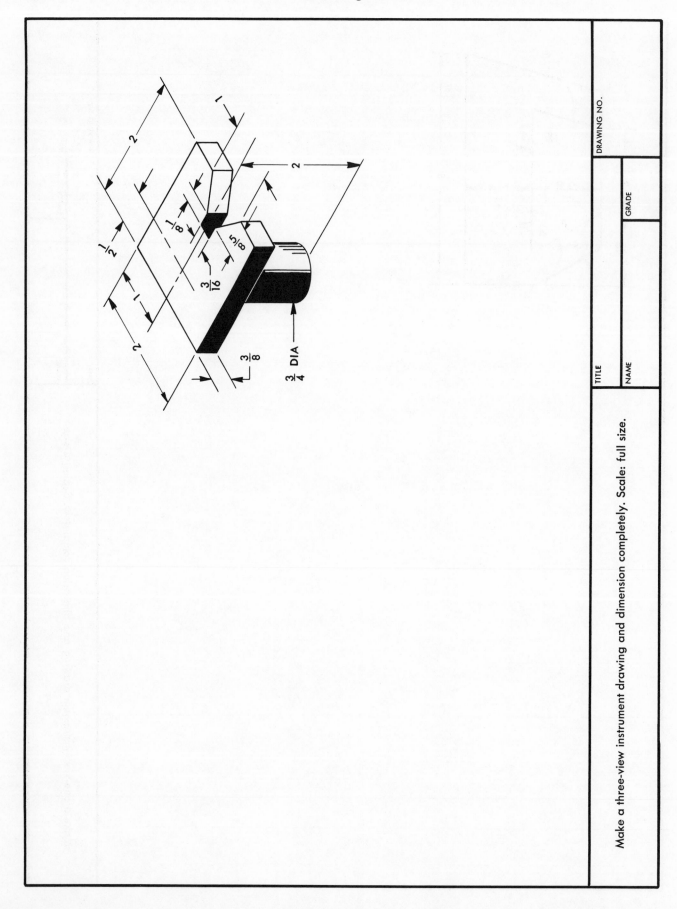

Make a three-view instrument drawing and dimension completely. Scale: full size.

DRAWING NO.

GRADE

TITLE

NAME

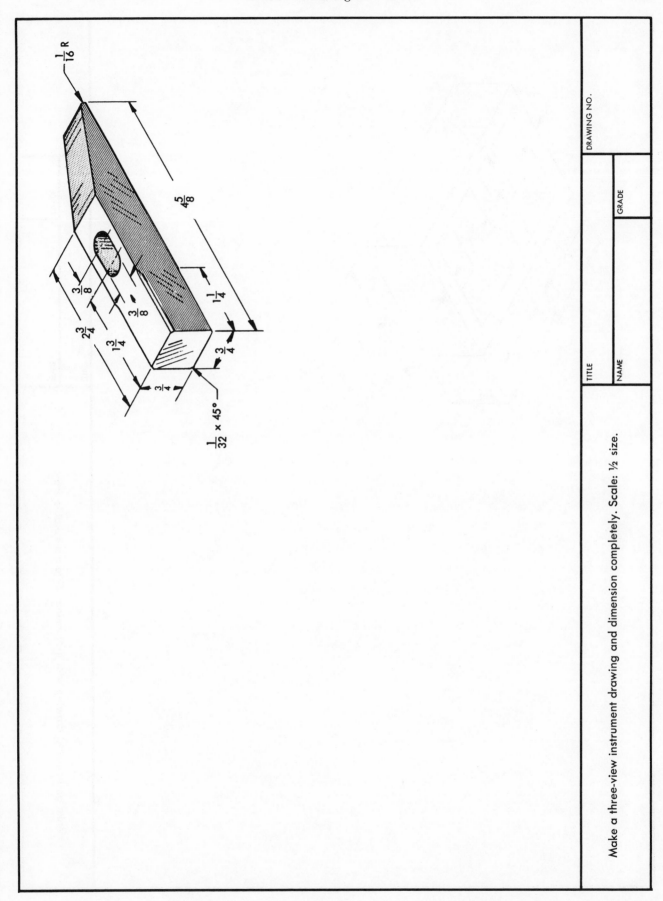

$\frac{1}{16}$ R

$4\frac{5}{8}$

$\frac{3}{8}$

$2\frac{3}{4}$

$\frac{3}{8}$

$1\frac{3}{4}$

$1\frac{1}{4}$

$\frac{3}{4}$

$\frac{3}{4}$

$\frac{1}{32} \times 45°$

DRAWING NO.

GRADE

TITLE

NAME

Make a three-view instrument drawing and dimension completely. Scale: ½ size.

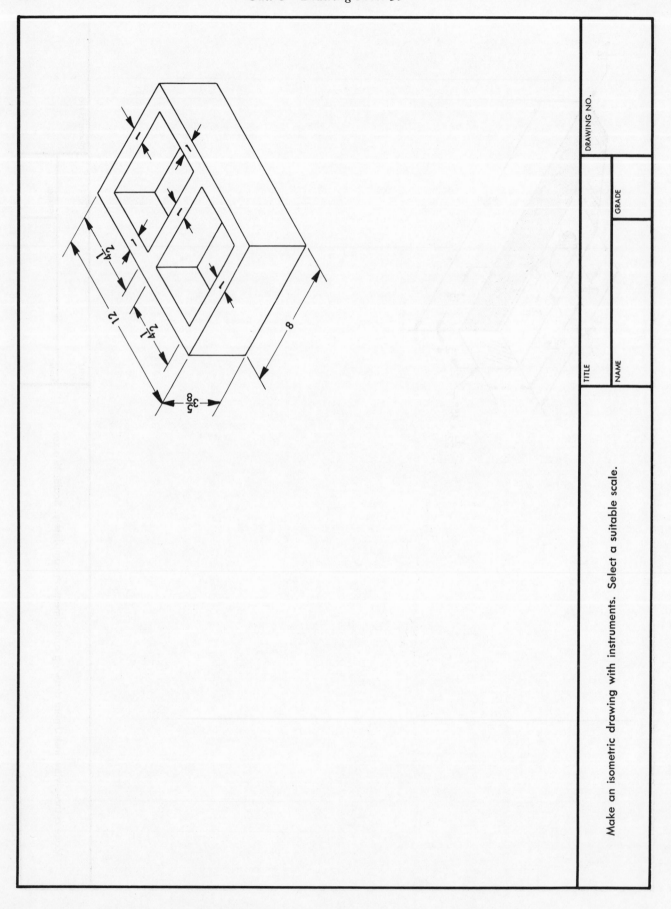

Make an isometric drawing with instruments. Select a suitable scale.

TITLE

NAME

GRADE

DRAWING NO.

Make an isometric drawing with instruments.
Scale: full size.

TITLE		DRAWING NO.
NAME	GRADE	

ALL STOCK 3/8 THICK

Make an oblique drawing using instruments. Select a suitable scale.	TITLE		DRAWING NO.
	NAME	GRADE	

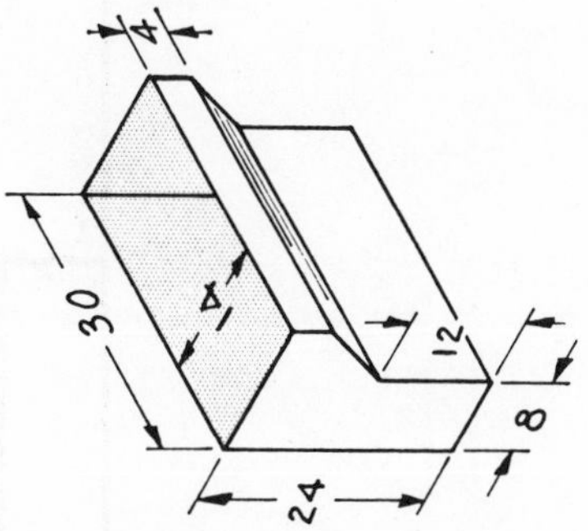

30

24

4

4

12

8

DRAWING NO.

GRADE

TITLE

NAME

Make a parallel perspective drawing using instruments. Select a suitable scale.

Make a three-view instrument drawing, with the front view a full section and dimension completely. Scale: full size.

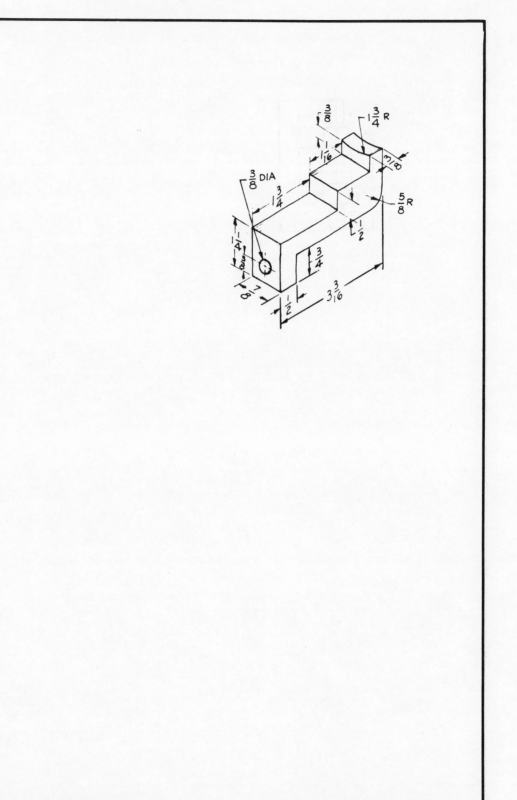

Make a two-view drawing with instruments. Scale: full size. Convert all fractional dimensions to decimal. Use conversion chart, inside back cover. Show tolerance of + and − .02 for hole.

TITLE		DRAWING NO.
NAME	GRADE	

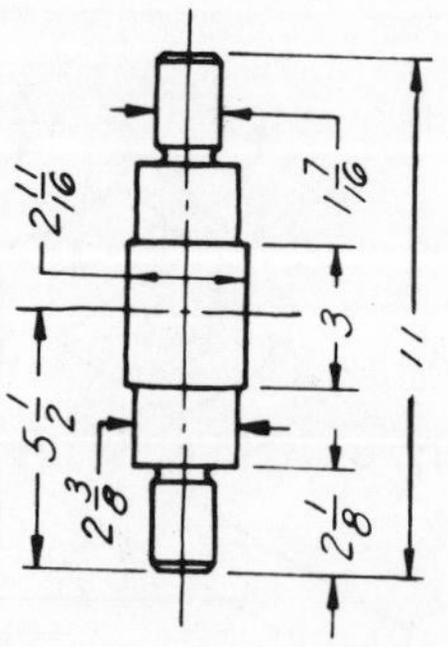

DRAWING NO.

GRADE

TITLE

NAME

Make a two-view drawing using instruments. Select a suitable scale. Convert all fractional sizes to decimals. Use decimal conversion chart, inside back cover. Include tolerance of + and −.005 for length only.

Drawing Geometric Constructions

In making a drawing there will be occasions when a knowledge of geometric construction, such as bisecting a line or arc, or drawing lines and arcs tangent to each other will be very helpful. Although the experienced draftsman uses many of these constructions, you will need only a few basic ones at this time.

Bisecting a line

1. To find the center of any given line such as AB, see Fig. 1, set the compass for any radius greater than one-half of AB. Using A and B as centers, draw two arcs to intersect at C, and two arcs to intersect at D.
2. Draw a straight line to connect points CD. The point where line CD crosses AB is the center of the line.

Bisecting an arc

1. To bisect arc AB of Fig. 2, set the compass at a radius greater than one-half of AB. With A and B as centers, draw two arcs intersecting at D and two arcs intersecting at E.
2. Draw a line connecting points D and E. The

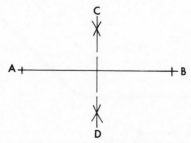

Fig. 1. Two pairs of arcs are used in bisecting a line.

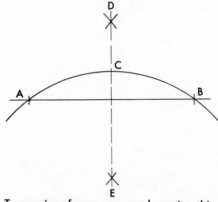

Fig. 2. Two pairs of arcs are used again, this time to bisect an arc.

point C where line DE crosses the arc represents the center of the given arc.

Bisecting an angle

1. To bisect an angle, see Fig. 3, use point A as a center and with the compass set at any convenient radius, draw an arc cutting line AB at D and line AC at E.
2. Set the compass at a radius greater than one-half of DE. With D and E as centers, draw two arcs to intersect at O.
3. Draw a line from A to O. The line AO bisects the angle.

Dividing a line into equal parts

1. Suppose you want to divide a given line into a number of equal parts such as eight. In some situations it may be difficult to use a scale to mark off equal divisions. In this case, the method of dividing the line into equal

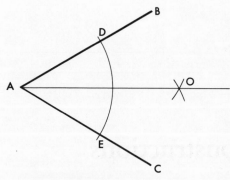

Fig. 3. Intersecting arcs are used in the procedure for bisecting an angle.

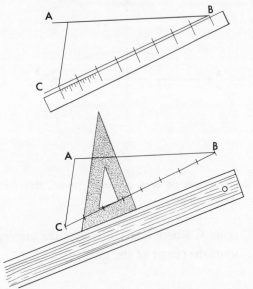

Fig. 4. A line can be divided into equal parts without using a scale.

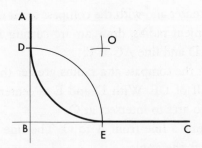

Fig. 5. The procedure for drawing an arc tangent to two lines at 90° is similar to the procedure for bisecting an angle.

parts by geometric construction is used. See Fig. 4. First draw line AB, which is the line to be divided.

2. Now draw line CB at any angle with AB. Starting at B on line CB, lay off eight equal

spaces either with dividers or a scale as illustrated.

3. From the point where the last space falls (C), draw a line connecting C with A. Then with a triangle, set parallel with line AC, draw lines from the points on line CB to line AB. The division points will be where the parallel lines intersect on line AB.

Drawing an arc tangent to two lines at 90°

1. Quite often you will need to draw an arc tangent to two lines as shown in Fig. 5. Draw the two lines AB and CB so that they are perpendicular and intersect at B.
2. Set the compass to the given radius and, with B as a center, draw an arc cutting line AB at D and line CB at E.
3. Then using the same radius and with D and E as centers, draw two arcs to intersect at O.
4. With O as a center and with the compass set at the same radius, draw the arc tangent to lines AB and CB.

Drawing an arc tangent to two lines not at 90°

1. When an arc must be drawn to two given lines making an angle of more or less than 90°, as seen in Fig. 6, set the compass at the given radius OR and using any points on lines AB and BC, draw a series of arcs.
2. Draw straight lines tangent to these arcs to

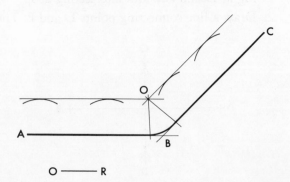

Fig. 6. Drawing an arc tangent to two lines not at 90°.

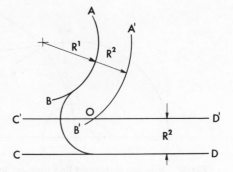

Fig. 7. Drawing an arc tangent to a straight line and an arc requires that the radii of both arcs be known.

form lines parallel to AB and BC and extend them to intersect at O.

3. With O as a center and the compass set at the given radius, draw the arc to lines AB and BC.

Drawing an arc tangent to a straight line and an arc

1. When it is necessary to draw an arc tangent to a straight line and an arc as shown in Fig. 7, set the compass to the radius R_1 of the given arc and draw arc AB. Draw the given straight line CD.

2. Set the compass a distance equal to R_1 plus the specified radius of the desired tangential arc R_2 and draw arc A'B'.

3. Draw line C'D' parallel to, and at a distance equal to R_2 from, line CD. Line C'D' will intersect arc A'B' at O.

4. With O as a center and the compass set at the given radius R_2, draw the arc tangent to arc AB and line CD.

Drawing tangent arcs

1. To connect arcs which are to be joined by a tangent arc of a given radius (R_2 as shown in Fig. 8) set the compass to R_1 and draw arc AB.

2. Set the compass to R_1 plus R_2, and draw the arc A'B'.

3. Set the compass to a radius equal to R_3 and draw the arc CD.

4. Set the compass to a radius equal to R_3 plus R_2 and draw arc C'D', to intersect arc A'B' at point O.

5. With point O as a center and the compass set at a radius equal to R_2, draw the arc BC tangent to arcs AB and CD.

Drawing a straight line tangent to two arcs

1. To connect arc AB and arc CD with a straight line as shown in Fig. 9, first find the difference between the radii of arcs AB and CD. Lay off this distance OE on the straight

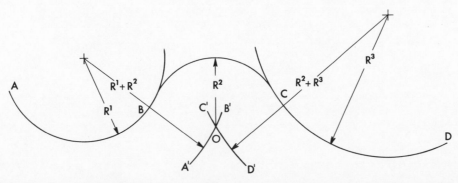

Fig. 8. When drawing an arc tangent to two adjoining arcs, first determine the radii of the adjoining arcs.

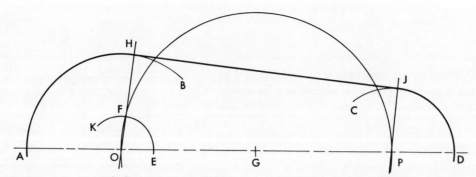

Fig. 9. Drawing a straight line tangent to two arcs is a common construction that you will use often.

line connecting the centers of arcs AB and CD. With O as a center and OE as a radius, draw the arc KE.

2. Now find the center of line OP. Then with OG as a radius and G as a center, draw arc OP, cutting arc KE at F.

3. Draw a line through points O and F, cutting arc AB at H. Now H becomes the point of tangency of arc AB.

4. Draw line PJ parallel with line OH. Point J is the point of tangency of arc CD.

5. Draw line HJ, which is the required tangent to the two arcs.

Draw the required geometric constructions on Drawing Sheets 42 and 43, pages 119 and 120. Follow the instructions at the bottom of each sheet.

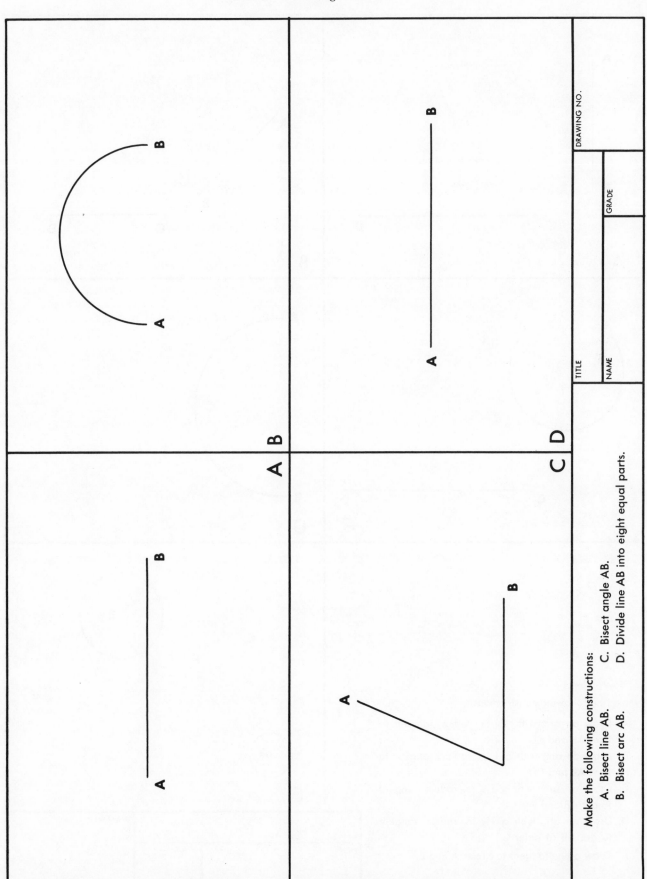

Make the following constructions:

A. Bisect line AB.   C. Bisect angle AB.
B. Bisect arc AB.   D. Divide line AB into eight equal parts.

DRAWING NO.

GRADE

TITLE

NAME

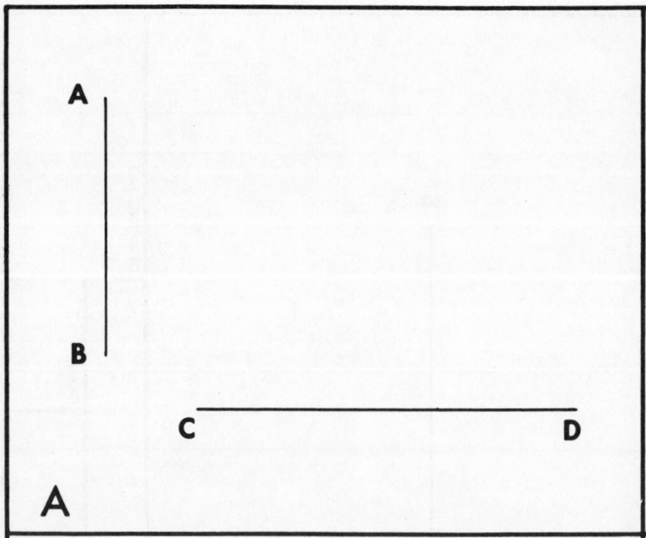

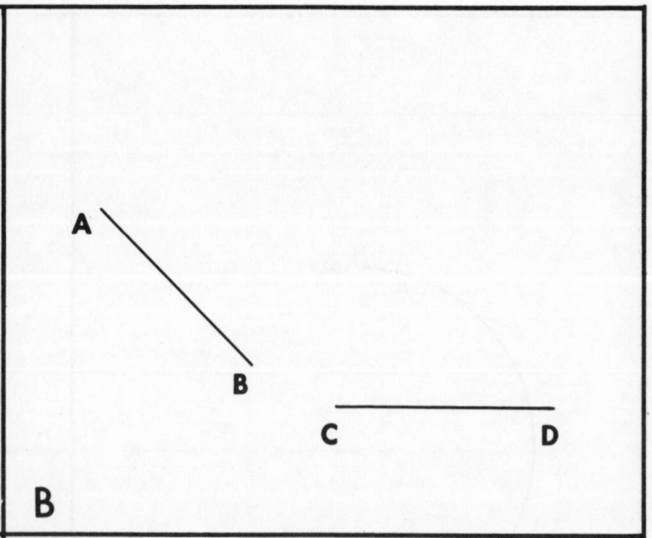

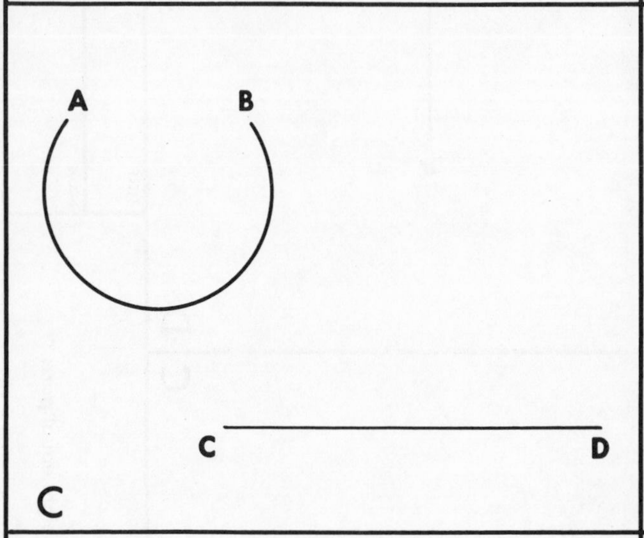

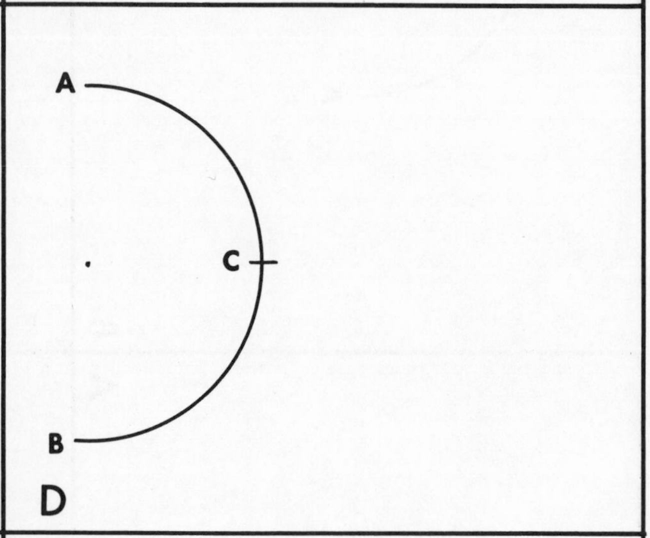

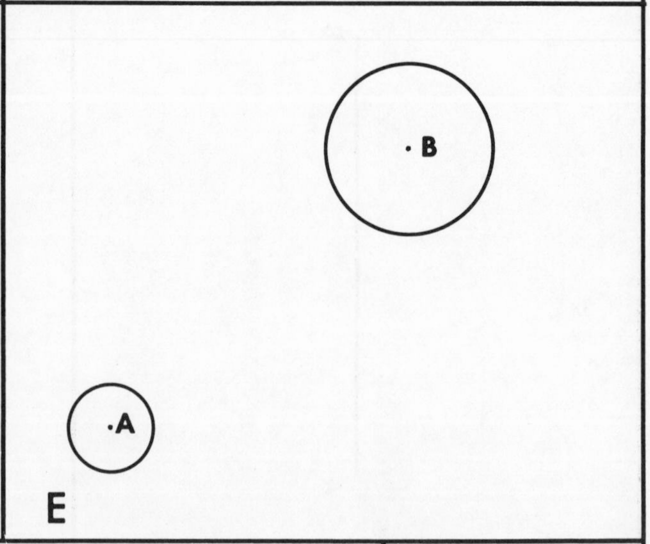

A

B

C

D

E

Make the following constructions:

A. Draw an arc, with 1¼ in. radius, tangent to lines AB and CD.

B. Draw an arc, with a 1 in. radius, tangent to lines AB and CD.

C. Draw an arc, with a 1⅛ in. radius, tangent to arc AB and line CD.

D. Draw an arc, with a 1¼ in. radius, tangent to arc AB at point C.

E. Draw a line tangent to circles A and B.

TITLE		DRAWING NO.
NAME	GRADE	

Architectural Drawing

In the future you may be involved in building a home. There are many agencies that provide the necessary advice as to how one should proceed in planning a home. However, you will be able to better plan the type of home you actually need if you understand a few basic principles of home planning and building. Furthermore, a knowledge of these principles will help you explain more clearly and intelligently to the architect just what you want.

Types of house architecture

Houses being built in our country today fall into two broad classifications: *traditional* and *modern*. *Traditional* homes are based on a definite architectural style which existed in Europe or America during some historic period. These, as you will see in Fig. 1, are referred to as English, Elizabethan, Georgian, Dutch Colonial, Southern Colonial, Cape Cod, French Provincial, Tudor, New England Colonial, and Spanish.

ENGLISH

GEORGIAN

ELIZABETHAN

DUTCH COLONIAL

Fig. 1. There are several basic types of architecture which are commonly referred to as traditional houses.

SOUTHERN COLONIAL

TUDOR

CAPE COD

NEW ENGLAND COLONIAL

FRENCH PROVINCIAL

SPANISH

Fig. 1. Continued.

The *modern style,* Fig. 2, differs radically from that of most conventional homes. Some have flat roofs with numerous windows. Others depart from square or rectangular outlines and have curved walls. In recent years, the Ranch and Split Level homes have become very popular.

Sketching preliminary floor plans

Once a person has decided on the type of home he can afford to build, he proceeds to make rough sketches of floor plans. The principal purpose of these floor plans is to show the general arrangement of the rooms, the size of the rooms, and other special features which one wants in the house.

In sketching preliminary floor plans, the use of squared or graph paper greatly simplifies the work. Each square of the paper represents a certain size, such as six inches or one foot. See Fig. 3.

Finished set of architectural plans

Once the rough sketches are completed, the next step is to consult an architect. The architect will probably make certain suggestions. When the details have been agreed upon, he will proceed to prepare a detailed set of plans. A detailed set of plans will usually include finished drawings showing a pictorial view of the house, a plot plan, a floor plan of the house, elevation views, sectional views, and detail drawings.

Fig. 2. These are the modernistic styles of homes being built at present.

RANCH

CONTEMPORARY

Architect—Frank Lloyd Wright

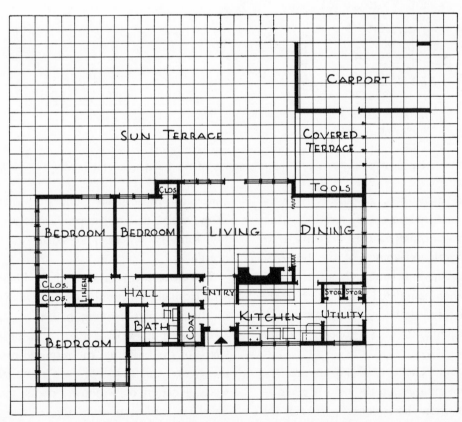

Fig. 3. Preliminary home planning is often done on graph paper to show the general arrangement of rooms.

Fig. 4. A perspective view gives the home owner a better idea of what the completed house will look like.

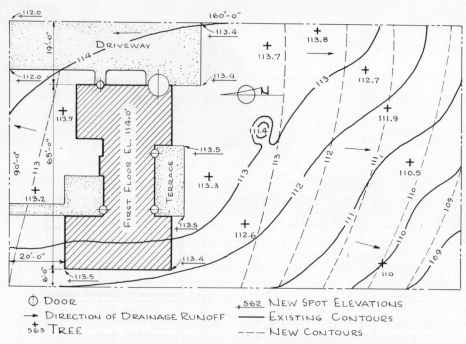

Ⓓ DOOR   ₊562 NEW SPOT ELEVATIONS
→ DIRECTION OF DRAINAGE RUNOFF   EXISTING CONTOURS
₊563 TREE   ---- NEW CONTOURS

Fig. 5. A plot plan shows the orientation of the building on the land.

The pictorial drawings. A perspective view of the house is featured on the pictorial drawing. See Fig. 4. The purpose of this drawing is to present the general appearance of the house so that the home owner can better visualize what it will look like when completed. In addition to the building, the pictorial drawing will often include other features such as trees, shrubbery, and the drive.

The plot plans. The overall layout and the orientation of the building on the land is given in the plot plan. It often shows contour lines of the ground to designate the grading changes which the contractor must make. The plot plan will also show walks, driveways, provisions for water supply, sanitary sewers, and storm water drainage. See Fig. 5.

The floor plan. A view of the building area as seen from above, cut along window level, is shown on the floor plan. See Fig. 6.

Included in a floor plan will be the outside shape of the building, the required room arrangement, size and shape of each room, thickness of walls, location of windows, doors, closets, stairs, fireplaces, and other auxiliary areas.

Dimensions are usually included outside and

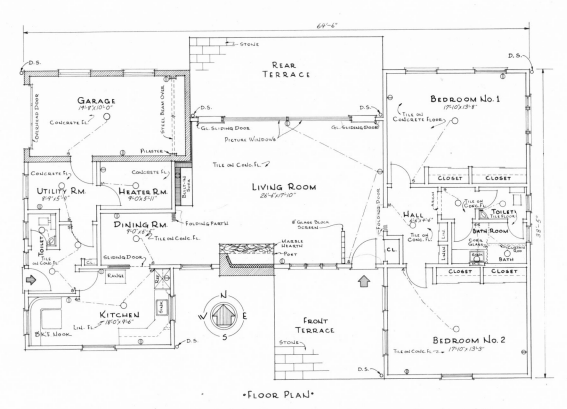

Fig. 6. This is a typical finished drawing of a floor plan prepared by the architect.

inside the building lines of plan views. Outside dimensions are limited to overall sizes, and sizes of openings in exterior walls. Inside dimensions locate sizes between the inside faces of building walls. Room sizes are usually given.

In preparing floor plans, the architect will use various symbols to show general construction details, such as exterior wall material and types of windows. Some of the more common symbols are illustrated in Fig. 7. If you understand these basic representations, you will find it easier to read prints of floor plans.

For small buildings, the practice is to include all of the necessary electrical, heating, and plumbing symbols on the floor plans. For large buildings, this information is shown on separate drawings.

The electrical system. Standard symbols are used on floor plans to show the various units of the electrical system, such as lights, outlets, switches, wire runs and other electrical components. Fig. 8 (Top) illustrates some of the more common symbols. Notice how these symbols are used in the floor plan shown in Fig. 6.

The heating and plumbing system. For residential buildings, the layout of the plumbing and heating systems are also usually incorporated in the floor plans. Separate piping and heating drawings are prepared for large structures. The residential floor plans will simply identify and show the location of the required units.

The plumbing contractor will determine the actual runs of pipes and make the necessary installations in relation to the placement of the specified fixtures. Similarly, the heating contractor will determine the actual duct system required for the heating units and make the installation accordingly. Plumbing and heating units are indicated on floor plans by means of symbols as shown in Fig. 8 (Bottom).

The elevation views. Elevation drawings show the exact shape, height and width of each wall, window, and door. Normally four views are included: front, right side, left side, and rear. These views are sometimes labelled north elevation, east elevation, west elevation, and south elevation. See Fig. 9.

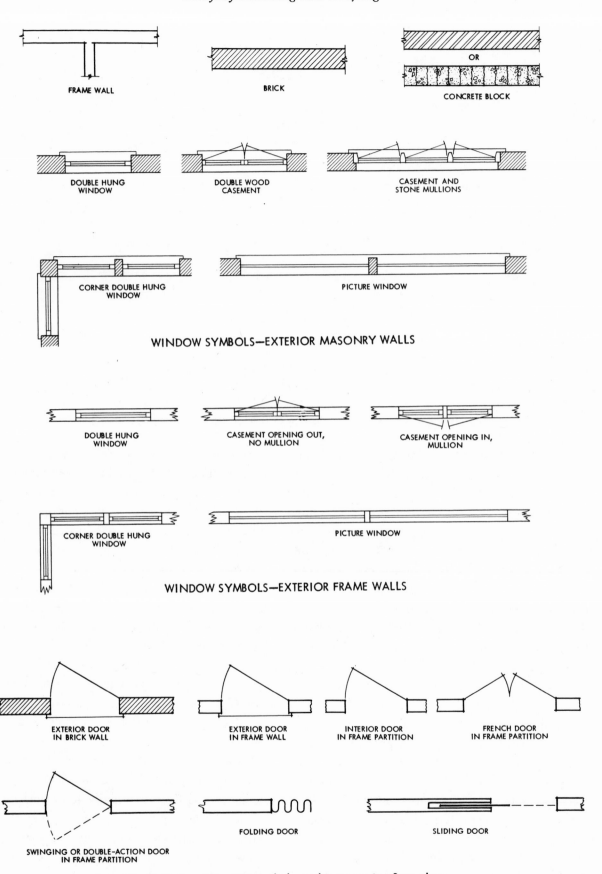

Fig. 7. Basic construction symbols used in preparing floor plans.

ELECTRICAL SYMBOLS

⒟	DROP CORD	S---	SINGLE-POLE SWITCH
	CEILING FIXTURE OUTLET	S³--	THREE-WAY SWITCH
	DUPLEX CONVENIENCE OUTLET		BELL
	WALL BRACKET		BUZZER
⊙	FLOOR OUTLET		TELEPHONE
	LIGHTING PANEL	Ⓕ	CEILING OUTLET FOR FAN
R	RANGE OUTLET	WP	WEATHERPROOF OUTLET

MECHANICAL SYMBOLS

HEATING SYMBOLS

12 - 3 - 38" RADIATOR SIZE - 12 COLUMN 3 SECTIONS - 38" HIGH	K E KITCHEN EXHAUST	SUPPLY DUCT	EXHAUST DUCT
WARM AIR INLET	COLD AIR RETURN	SECOND FLOOR SUPPLY	SECOND FLOOR RETURN

PLUMBING SYMBOLS

OR TOILET OR WATER CLOSET	RECESSED BATH TUB	PLAIN SINK	L | T DOUBLE LAUNDRY TRAY
LAVATORY USE SPECIFICATIONS TO DESCRIBE	SHOWER STALL	F D FLOOR DRAIN	W H WATER HEATER

Fig. 8. These are the most common electrical, heating, and plumbing symbols used on finished floor plans.

Very few dimensions are shown on elevations. Height and locations of windows, ceiling, etc., are specified. Usually no horizontal dimensions are included. Most essential sizes are placed either on the floor plan or sectional views.

Solid lines are used almost entirely in elevation views. Dash lines are used only to show the foundation of the building.

Elevation views will generally show the building materials and other special construction features. These details are represented by symbols which are fairly well standardized and recognized by architects, draftsmen, contractors, and tradesmen. See Fig. 10.

The sectional views. Parts of the building structure which are common throughout the build-

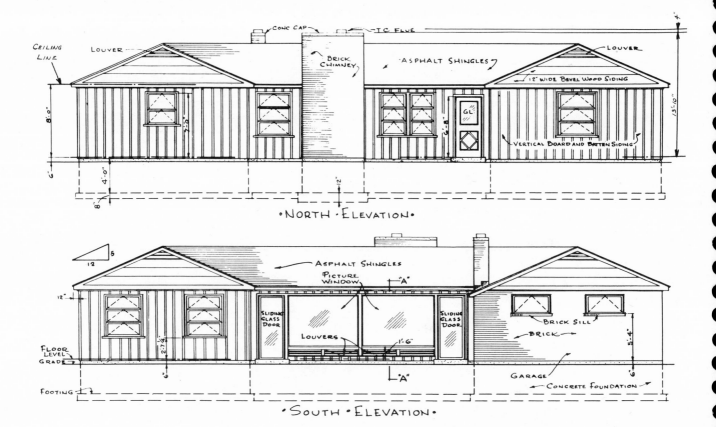

·NORTH·ELEVATION·

·SOUTH·ELEVATION·

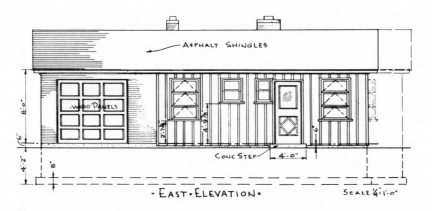

·EAST·ELEVATION·

SCALE ¼"=1'-0"

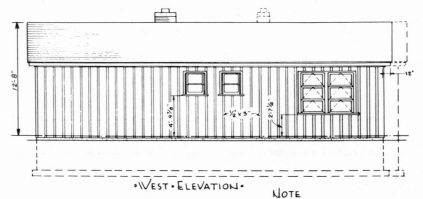

·WEST·ELEVATION·

NOTE
WINDOW SILLS SHALL BE REDWOOD
1x10 VT&G SHALL BE REDWOOD OR CEDAR.
ALL OTHER EXPOSED EXTERIOR WOOD SHALL BE PINE.
ALL EXT. WOOD SHALL BE SURFACED

Fig. 9. Typical elevation drawings show all sides of the building.

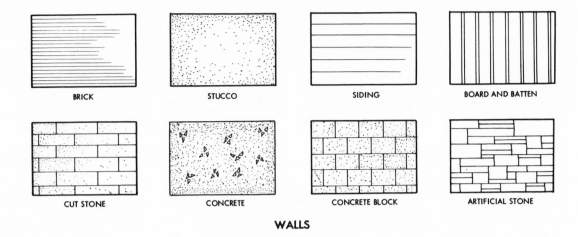

WALLS

DOORS

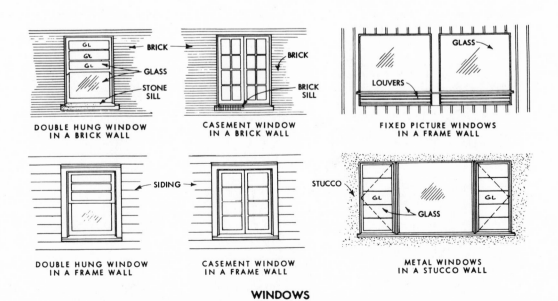

WINDOWS

Fig. 10. The most common symbols used in elevation drawings.

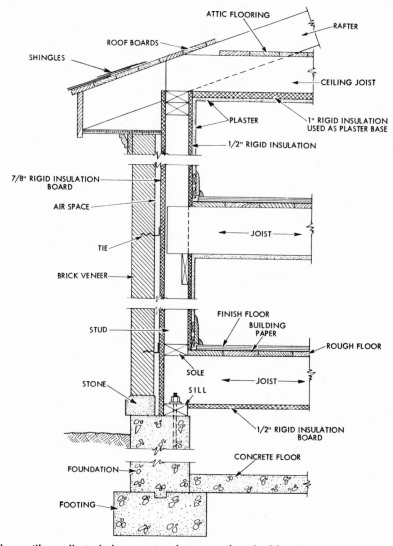

Fig. 11. Building plans will usually include a sectional view to show building features common to the entire house.

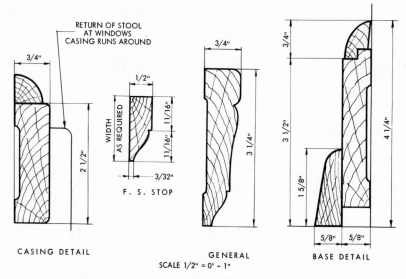

Fig. 12. Detail drawings are often included in a set of building plans to show how certain features are to be constructed.

ing and which reflect repeatedly used construction features are shown in sectional views. See Fig. 11.

The detail drawings. Often drawings are necessary to show how certain specific items are to be constructed or placed. All items such as these are shown on detail drawings. Thus a detail may present the construction method required for doors, windows, eaves, cabinets, and similar items. These drawings are used whenever the information provided in elevation, floor plan, or sectional views

fixtures, interior and exterior wall finishes, grading, and other building features. See Fig. 14.

The detailed specifications give the owner definite assurance that the building project will be completed in accordance with previous agreements. The building specifications also give the contractor and sub-contractor a clear picture of what they are bidding on and what installations will be required of them during the construction stages.

CODE	QUAN	TYPE	DOOR SIZE	MFRS No	REMARKS
Ⓐ	2	WD, 1 PANEL, 1 GL	3'-0"×6'-8"×1⅜"	3070D	SEE ELEVATION VIEWS FOR PANEL DESIGN
Ⓑ	2	GL SLIDING	3'-0"×7'-0"×1"	3070G	1" THERMOGLAZE GL
Ⓒ	1	OVERHEAD	7'-0"×8'-0"	7080	SEE ELEVATION VIEW FOR PANEL DESIGN
Ⓓ	2	WD, HOLLOW CORE *	2'-10"×6'-8"×1⅜"	21068	BIRCH
Ⓔ	5	WD, HOLLOW CORE * SLIDING	2'-8"×6'-8"×1⅜"	2868	"
Ⓕ	3	WD, HOLLOW CORE *	2'-6"×6'-8"×1⅜"	2668	"
Ⓖ	3	" " " *	2'-4"×6'-8×1⅜"	2468	"
Ⓗ	4	" " " *	3'-0"×6'-8"×1⅜"	3068	"
Ⓙ	2	" " " *	2'-0"×6'-8"×1⅜"	—	"
Ⓘ	1	WD, HOLLOW CORE *	2'-8"×6'-8"×1⅜"	2868	"

·DOOR·SCHEDULE·

* FLUSH DOOR

CODE	QUAN	TYPE	SASH SIZE	MFRS No	REMARKS
Ⓚ	7	AWNING 3 LTS	2'-9⅞"×4'-6"	66	SEE SPEC FOR GL
Ⓛ	3	" 3 LTS	3'-9"×4'-3¾"	1010	" " " "
Ⓜ	1	" 2 LTS	3'-4"×3'-4"	86	" " " "
Ⓝ	3	D H 2 LTS	1'-7"×2'-0⅝"	2	OBSCURE GL SEE SPEC
Ⓞ	1	D H 2 LTS	1'-1"×2'-0⅝"	3	SEE SPEC FOR GL
Ⓟ	2	AWNING 1 LT	3'-9"×1'-6"	10	" · " "
Ⓠ	2	FIXED 1 LT	8'-0"×5'-5"	918	1" THERMOGLAZE GL
Ⓡ	1	AWNING 3 LTS	3'-3"×4'-6"	1003	SEE SPEC FOR GL
Ⓢ	1	" 3 LTS	2'-9"×4'-3"	1007	" " " "

·WINDOW·SCHEDULE·

Fig. 13. A typical window and door schedule, which is usually found on the floor plan or elevation drawing.

is not sufficiently clear for the workers to follow. See Fig. 12.

The window and door schedules. To avoid cluttering drawings with unnecessary dimensions, windows and doors on drawings are coded with letters enclosed in circles. The same letters are used for all windows and doors when they are of identical construction and size. The coded letters and titles are then shown in a schedule similar to the one in Fig. 13. The window and door schedule is placed on the elevation drawing, the floor plan, or on a separate sheet.

Building specifications

A building specifications list is an outline prepared by the architect which spells out all of the construction details which cannot be shown on the set of building plans. This includes the kind of materials to be used, quality of workmanship, type of heating, ventilating, electrical, and plumbing

Sec. 4. GRADING   3

a. Grading. Do all cutting, backfilling, filling, and grading necessary to bring all areas within property lines to the following subgrade levels:

   (1) For paving, walks, and other paved areas to the underside of the respective installation as fixed by the finished grades therefor.

   (2) For lawns and planted areas to four (4) inches below finished grades.

b. Material for Backfill and Fill. All material used for backfill and fill shall be free from deleterious materials subject to termite attack, rot, decay or corrosion, and frozen lumps or objects which would prevent solid compaction.

   Materials for backfill and fill in various locations shall be as follows:

   (1) For Interior of Building. Sand or an approved properly graded mixture of sand and gravel. Foundry sand shall not be used.

   (2) For Exterior Under Paving. Use excavated materials free from top soil, or other materials approved by the Architect.

   (3) For Use Under Lawns and Planted Areas. Use, after Architect's approval, excavated materials with admixture of top soil or earth. Heavy clay shall not be used.

c. Backfill Against Foundation Walls shall:

   (1) Be done only after work to be concealed thereby has been inspected and approved by the Architect.

   (2) Be deposited in six (6) inch layers, each to be solidly compacted by tamping and puddling.

d. Subgrades for Lawn and Planted Areas. Slope the subgrade evenly to provide drainage away from building in all directions at a grade of at least 1/4 in. per ft.

e. Settlement of Fills. Fill to required subgrade levels any areas where settlement occurs.

Fig. 14. The list of specifications tells the owner exactly what will be installed in his new home.

Check your knowledge of this unit by completing Self-Check for Unit 10, Parts 1, 2, 3, 4 and 5, pages 133-141.

Complete the sketch on Drawing Sheet 44, page 142. Follow the instructions on the bottom of the sheet.

SELF-CHECK FOR UNIT 10

PART 1

DIRECTIONS: Circle the letter T if the statement is True or the letter F if the statement is False.

Self-Check answers may be found on page 172.

1. T F The first step in the design and planning of a home is usually a rough floor plan.

2. T F The architect usually provides a pictorial view of the house so that the owner will have a clearer idea of the design.

3. T F The plot plan usually shows four walls and the roof of the building.

4. T F Electrical, heating and plumbing symbols are found on the architect's floor plan.

5. T F Elevation drawings usually show four views of the building.

6. T F Elevation views never show the types of building materials, height or location of windows and doors.

7. T F A sectional view will show a section of the building wall and is used to show conventional construction features of the house.

8. T F For special construction features, such as the fireplace, cabinets, or a staircase, a detail drawing is usually provided.

9. T F The window and door schedule is usually found in the plot plan.

10. T F Building specifications spell out all of the construction details which cannot be shown on the set of building plans.

11. T F The building specifications are useful to the contractor, as well as to the owner.

12. T F The size of the lot upon which the home is to be built may be found in the plot plan.

13. T F The floor plan would not include types of exterior finishing materials.

14. T F The architect will usually want the owner to provide a rough floor plan.

15. T F The plumbing will normally be installed by a subcontractor.

16. T F A complete set of building plans is the rough floor plan, a pictorial view, elevation views, a plot plan, detail drawings, and a building specifications list.

17. T F The plot plan shows contour changes to be made on the lot.

18. T F A joist is a vertical beam which supports the foundation.

134

19. T F Eaves provide structural support for the floors and ceiling.

20. T F An architectural set of plans will contain one or more typical sections showing complete construction details of common building features.

Score_____

PART 2

DIRECTIONS: Identify the symbols which appear below. Write the name for each in the space provided. *Self-Check answers may be found on page 172.*

1. _____

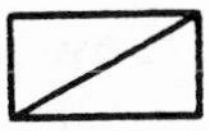

2. _____

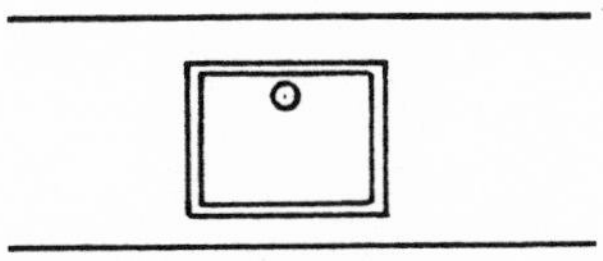

3. _____

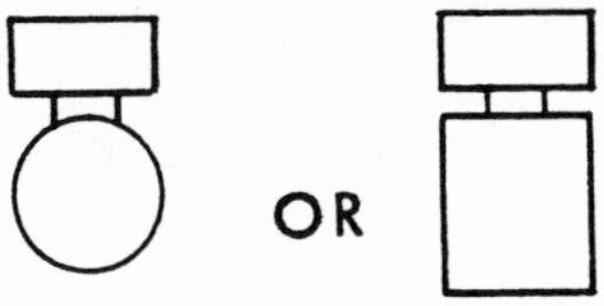

4. _____

5. _____

6. _____

7. _____

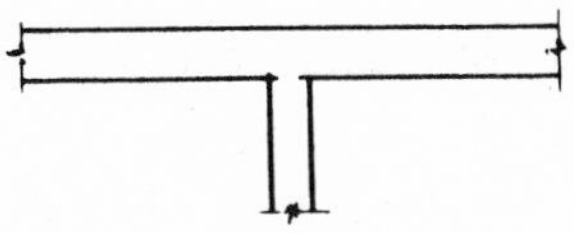

8. _____

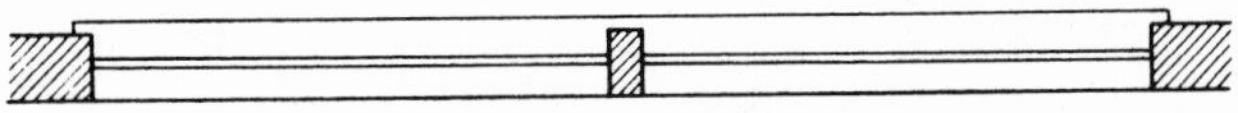

9. _____

Score_____

PART 3

DIRECTIONS: Using Print 1, answer the questions listed below.

Self-Check answers may be found on page 172.

Questions	Answers
1. What are the over-all dimensions of the house?	1. _____
2. How many windows in the floor plan?	2. _____
3. How many closets are shown, including linen closets?	3. _____
4. From this first floor plan, how can you tell that there is a basement?	4. _____
5. What is the size of the living room?	5. _____
6. How many closets are shown opening into the bedrooms?	6. _____
7. How many windows are shown in the two rear bedrooms?	7. _____
8. How many doors are shown in the kitchen?	8. _____
9. What is the size of the kitchen?	9. _____
10. How many exterior doors, including garage, are shown?	10. _____
11. How many windows are shown in the kitchen?	11. _____
12. What is the size of the family room?	12. _____

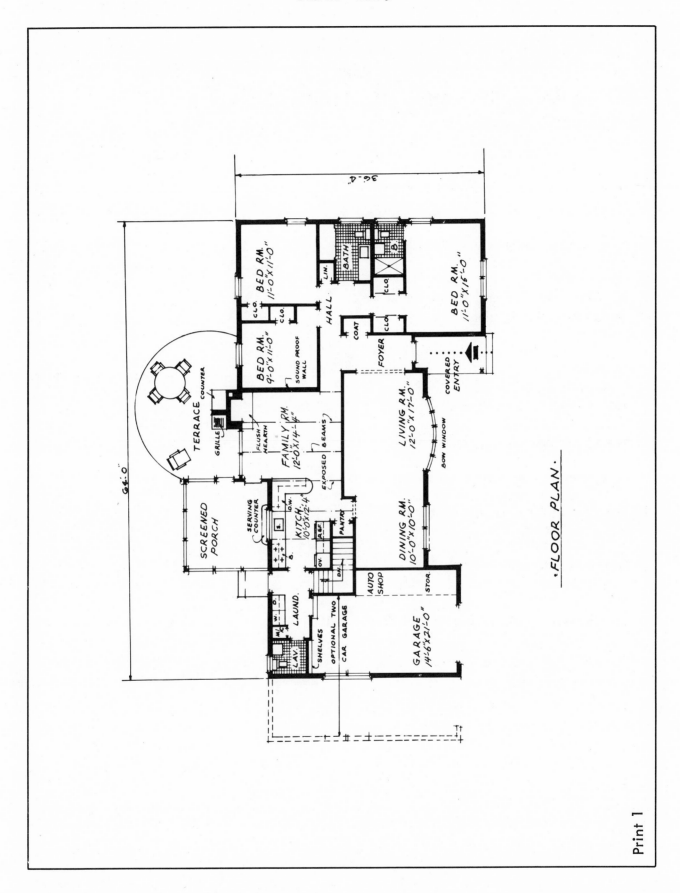

·FLOOR PLAN·

138

PART 4

DIRECTIONS: Using Print 2, answer the questions listed below.

Self-Check answers may be found on page 172.

Questions

Answers

1. How many doors and windows are shown in the south elevations?

1. _____

2. How far does the terra cotta flue lining extend above the brick work?

2. _____

3. How far above the floor line is the top of the door in the north elevation?

3. _____

4. What type of doors are shown in the south elevation?

4. _____

5. How many louvers are shown?

5. _____

6. What is the height of the ceiling?

6. _____

7. How deep is the foundation and footing?

7. _____

8. What material is to be used in the construction of the chimney?

8. _____

9. How far do the eaves overhang the sides of the building?

9. _____

10. What material is used to cap the chimneys?

10. _____

Score_____

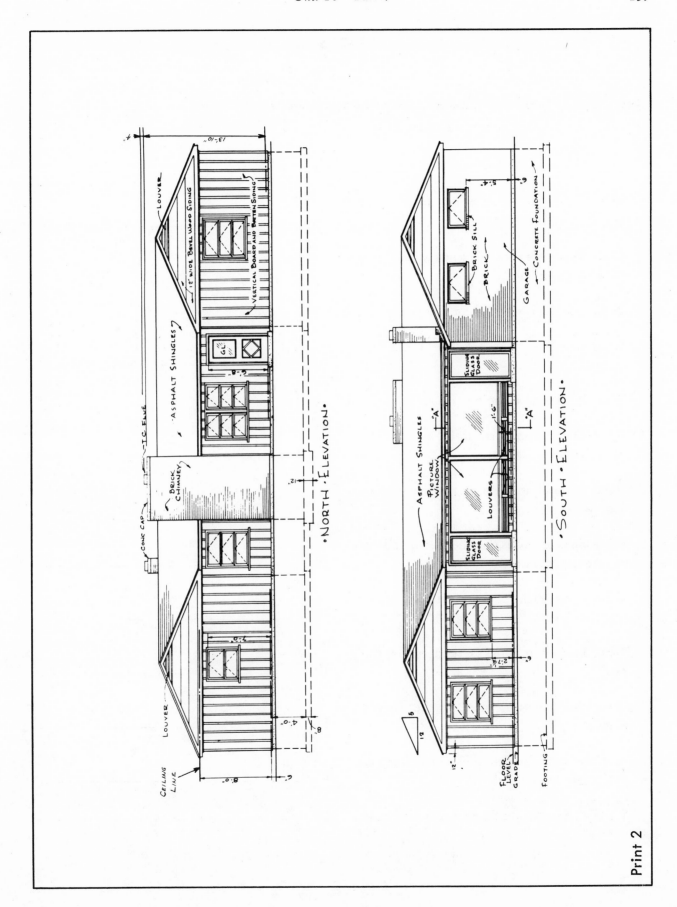

140

PART 5

DIRECTIONS: Using Print 3, answer the questions listed below.

Self-Check answers may be found on page 172.

Questions   Answers

1. What is the distance from the floor line to the top of the roof?

1. _____

2. What is the height of the center windows in the west elevation?

2. _____

3. What material is specified for the garage?

3. _____

4. What kind of wood is indicated for the window sills?

4. _____

5. Of what material is the step made?

5. _____

6. What kind of panels are to be used for the garage door?

6. _____

7. What material is specified for the exterior west wall?

7. _____

8. What kinds of woods shall be used on all exposed exterior surfaces?

8. _____

9. What is the height of the step shown in the east elevation?

9. _____

10. What kind of material is specified for the roof?

10. _____

11. What size battens are used for the exterior siding?

11. _____

12. How far above the floor line are the bottoms of the three-pane windows?

12. _____

13. What scale is used for the elevation drawings?

13. _____

Score_____

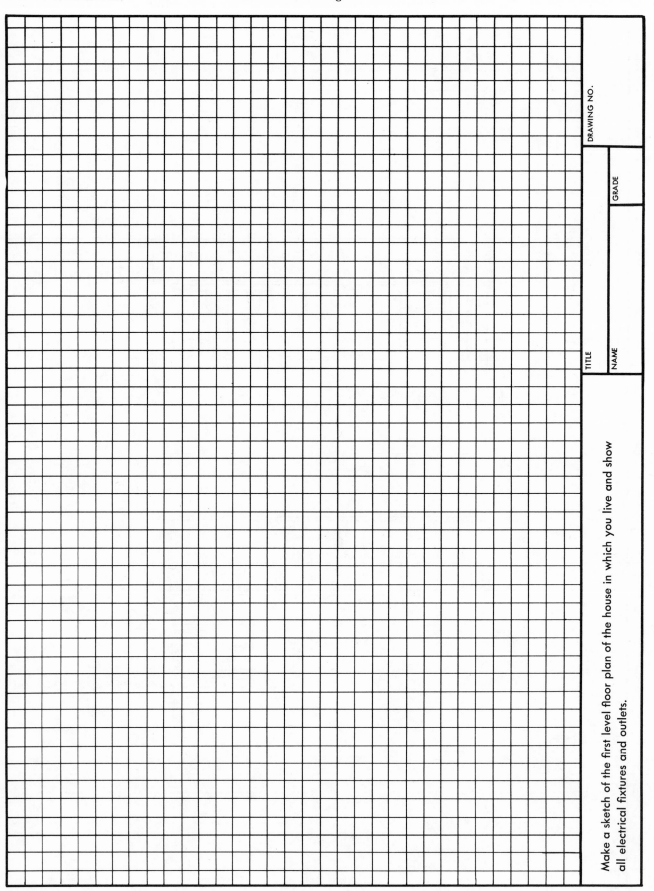

DRAWING NO.

TITLE

NAME   GRADE

Make a sketch of the first level floor plan of the house in which you live and show all electrical fixtures and outlets.

Drawing Charts

Charts are graphical aids used in presenting various kinds of information so it is more readily understood by the reader. Thus, by means of a chart, data which otherwise would be somewhat confusing becomes much more meaningful and significant. Almost any newspaper today will contain charts of one kind or another. Charts are used extensively by the scientist, engineer, economist, statistician, accountant, and others as a rapid and simplified means of presenting material to someone who is not familiar with the subject.

There are many varieties of charts, each of which is designed for a particular function. In the past many charts were often referred to as graphs. The more popular practice today is to call all such graphic aids charts. Several of the more common types of charts are described in this unit.

Line charts

A line chart is intended to illustrate the relationship between two sets of data. The method of presenting data on a line chart begins with the construction of two perpendicular lines, intersecting at a point called the origin. See Fig. 1. The horizontal line is referred to as the *X-axis* or *abscissa*. The vertical line is referred to as the *Y-axis* or *ordinate*. The X and Y axes are referred to as coordinate axes. The coordinate axes separate the four quadrants which are numbered I, II, III, IV. Most charts, however, only use the first quadrant.

The two sets of data are plotted along the two axes. The numbers that tend to fluctuate, or are irregular, are plotted along the vertical axis; the numbers that are regular, or that increase or decrease at regular intervals, are usually plotted along the horizontal axis.

Convenient scales are first selected for the two sets of data, then the units of the two axes are marked according to the scales selected. The data is then plotted on the graph and lines are drawn connecting the plotted points.

In Fig. 2, the distance required to stop a car is

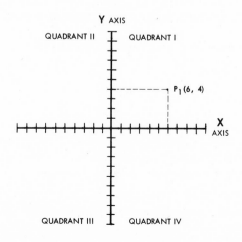

Fig. 1. A point is plotted on a graph by finding the intersection of its X and Y values.

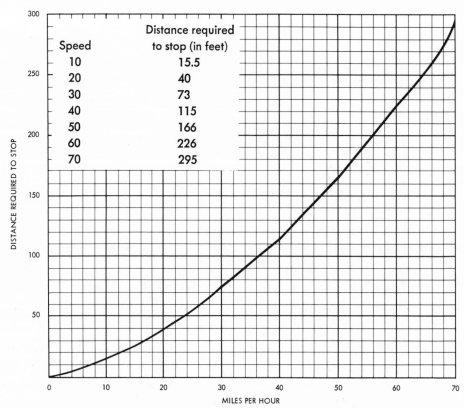

Fig. 2. A line chart shows the relationship between two sets of data.

plotted in relation to the speed of the car. The two sets of data are the distance, in feet, required to stop the car, and the speed of the car. The speed of the car represents the more regular, or easily controlled, of the two sets of data. The speed will be plotted along the horizontal axis. The stopping distance is the more variable of the two sets of data, and will be plotted along the vertical axis.

Examining the two sets of data, it is seen that the speed ranges from 10 mph to 70 mph. Therefore the horizontal axis will be scaled from 0 to 70 mph, each unit along the X-axis representing

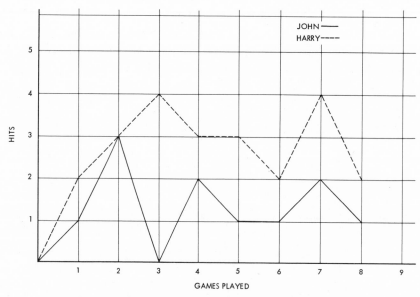

Fig. 3. Several sets of figures can be compared on one chart.

10 mph. The stopping distance ranges from 15.5 to 295 feet. The vertical scale therefore will range from 0 to 300 feet. Each unit along the Y-axis will represent 50 feet.

The points are plotted by consulting the two sets of data. At 10 mph the stopping distance for the car is 15.5 feet. Placing a pencil along the 10 mph line, glance over to the vertical axis which has been scaled in units of 50 feet. Each large unit contains 5 smaller units which represent 10 feet. Counting up the 10 mph line to 1½ small units (or 10 feet plus 5 feet), mark the point with the pencil. Using the next pair of figures, 40 feet at 20 mph, repeat the above procedure. Continue plotting the pairs of figures until all the points are plotted. Finally, connect all the points with a line.

In each chart problem, always figure out the facts to be represented, range of the facts or numbers, and the value to be assigned for each unit on both axes before attempting to lay out the chart. Be sure there is a title that is brief and adds to the meaning of the chart.

A line chart is often used to show a comparison between two or more sets of facts. Notice in Fig. 3 how it is possible to compare the number of hits you made in several ball games with those of some other player on the team. First plot your hits on the sheet. Then do the same for the hits made by the other player. Use a solid line to represent your hits and a dotted line for those of the other player. See how easy it is to compare the two records.

Keep in mind that when several sets of facts are plotted on one sheet, each chart line must be displayed differently if the comparison is to stand out clearly. The practice is to use solid and dotted lines of different weights or to make the lines a different color.

Bar charts

The bar chart is so called because of the heavy bars which appear on the chart to represent the proportionate amount of a numerical value. These bars should be made to start from zero and may be placed horizontally or vertically. The scale should be the same for each bar and should be lettered

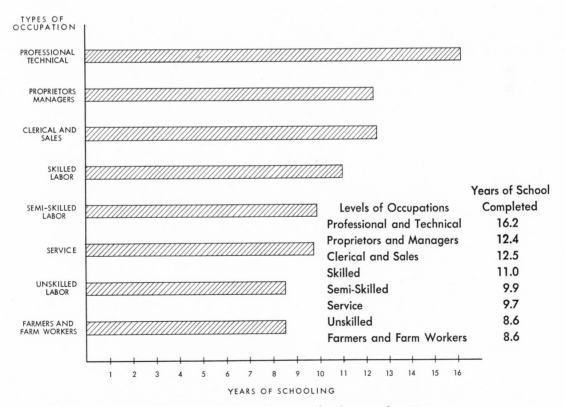

Levels of Occupations	Years of School Completed
Professional and Technical	16.2
Proprietors and Managers	12.4
Clerical and Sales	12.5
Skilled	11.0
Semi-Skilled	9.9
Service	9.7
Unskilled	8.6
Farmers and Farm Workers	8.6

Fig. 4. A bar chart may have the bars in a vertical or horizontal position.

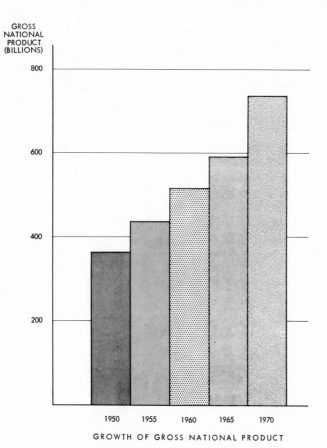

Year	Total GNP (in Billions)
1950	352.2
1955	434.9
1960	505.0
1965	598.0
1970	731.7 (Projected)

Fig. 5. An example of a bar chart with bars in a vertical position.

GROWTH OF GROSS NATIONAL PRODUCT

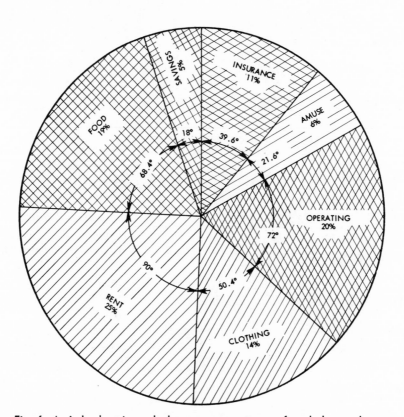

Fig. 6. A circle chart is used whenever percentages of a whole are shown.

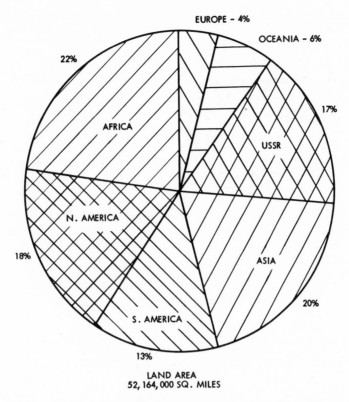

EUROPE - 4%

OCEANIA - 6%

22%

17%

AFRICA

USSR

N. AMERICA

18%

ASIA

20%

S. AMERICA

13%

LAND AREA
52,164,000 SQ. MILES

Fig. 7. With a circle chart, proportions are clearly and quickly seen.

along the bottom or left side of the chart to permit reading the numerical values. The name of the items represented should be lettered on the appropriate axis and a title should be included.

In Fig. 4, the educational levels of people at various levels of employment are compared. Notice that in constructing the chart, data is arranged exactly as it would be were the chart a line graph. The constant or controlled data, years of schooling, is plotted along the horizontal axis. The variable data, the different levels of occupation, is plotted, or arranged, along the vertical axis. Notice that those in the higher level positions have correspondingly higher educations.

Fig. 5 represents another type of bar chart. The bars are vertical rather than horizontal. The bars are not spaced as in Fig. 4, and are cross-hatched in order to present a more distinct appearance and to more easily read the chart. The chart illustrates the growth of the gross national product, a measure of the goods and services sold in this country, at five year intervals for a period of twenty years.

Circle charts

The circle, or pie chart is used frequently because it is so easy to construct. Its chief function is to show a graphic comparison of related quantities expressed in percentages. This chart is in the form of a circle. The total area of the circle represents 100 per cent of the whole being illustrated. The various parts of the circle, or sectors, are easy to calculate. They are a percentage, or proportionate amount, of the entire circle. See Fig. 6.

To determine the number of degrees in each of the sectors, multiply the percentage of the whole that the sector represents by 3.6. For example, if food were 25 per cent of the total operating expenses in a chart similar to Fig. 6, the number of degrees in the sector that represents the amount spent for food would be found by multiplying 25 by 3.6. In this case, 25 × 3.6 = 90 degrees.

Once the size of all the sectors has been determined, the chart may be drawn. Draw the circle and then measure the divisions using a protractor to measure the calculated degrees. Describe the nature of each quantity and its percentage. Complete the chart by lettering in the title.

Fig. 7 illustrates another example of a circle chart. In this chart, the total land area of our planet is shown divided into the familiar land masses. Notice that the USSR is almost as large as Europe and South America combined.

Check your knowledge of this unit by completing the Self-Check for Unit 11 on pages 149 and 150.

Complete Drawing Sheets 45, 46, 47, and 48, pages 151-154. Follow the instructions on the bottom of each sheet.

SELF-CHECK FOR UNIT 11

PART 1

DIRECTIONS: Complete the following statements by writing the correct word or words in the blank spaces provided.

1. The most frequently used types of charts are _____, and _____.

2. The horizontal and vertical lines in a line chart are referred to as the _____ axis and the _____ axis.

3. In a line chart, the set of data that is constant or regular is plotted along the _____ axis.

4. The set of data that is irregular and fluctuates is plotted along the _____ axis.

5. The _____ of data must be taken into consideration when determining the numerical value of each unit along the axes.

6. If a person's bowling average over a period of weeks was plotted on a line graph, the _____ line would be the bowling average line.

7. In a bar chart, the bars can be arranged either in a _____ or _____ position.

8. If 90° of a circle chart represents the amount spent for clothing, clothing must be _____ per cent of the budget.

9. In a circle chart, 50 per cent of the whole would equal _____ degrees.

10. The easiest type of chart to construct is the _____ chart.

PART 2

DIRECTIONS: Circle the letter T if the statement is True or the letter F if the statement is False.

1. T F Charts are used in presenting facts and figures in a concise manner.

2. T F Before marking the values of the units along the axes, the range of the data must be considered.

3. T F A sector of a circle chart representing 25% would contain 60°.

4. T F The best chart to use in showing our national income over a period of ten years would be a circle chart.

5. T F The title of a chart should be very detailed.

6. T F Both bar graphs and line graphs can be used to show differences in amounts over a period of time.

7. T F Charts are sometimes referred to as graphs.

8. T F A sector of a circle containing 18° would represent a value of 5%.

9. T F In making a line chart showing the daily temperature readings, the horizontal axis would be used for the readings.

10. T F If over half of your weekly allowance is spent for recreation, a section containing more than 90° but less than 180° could be used to show this on a circle chart.

150

PART 3

What per cent of the whole does each of the sectors of this pie chart represent?

1. _____

2. _____

3. _____

4. _____

5. _____

6. _____

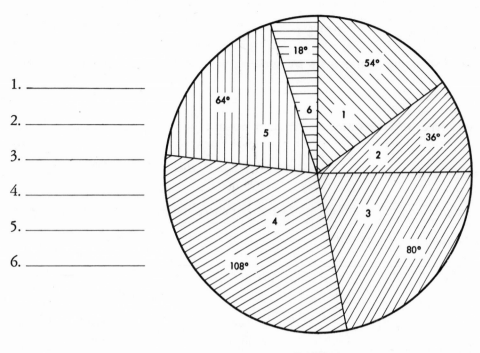

PART 4

Construct a bar chart with six equally spaced vertical bars, representing values of 30%, 50%, 60%, 75%, 80%, and 95%.

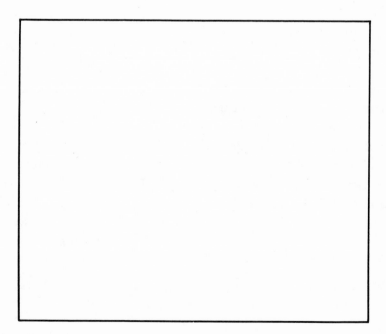

Score_____

Temperature	Hour
60°	7 A.M.
65°	8 A.M.
68°	9 A.M.
70°	10 A.M.
74°	11 A.M.
80°	12 P.M.
84°	1 P.M.
86°	2 P.M.
90°	3 P.M.
70°	4 P.M.
68°	5 P.M.
60°	6 P.M.
55°	7 P.M.

Construct a line chart showing the above hourly temperatures. Use any desired units.

TITLE

NAME

GRADE

DRAWING NO.

Hole	Strokes
1	4
2	5
3	6
4	3
5	7
6	4
7	4
8	5
9	6

Hole	Strokes
10	3
11	4
12	8
13	7
14	4
15	4
16	5
17	6
18	5

Assume that you played eighteen holes of golf. Construct a vertical bar chart using the given data:

TITLE

NAME

GRADE

DRAWING NO.

Size of
Common Nail   Length of nail (in inches)
    6d   2
    8d   2½
    8d   2¾
    10d   3
    12d   3¼
    16d   3½
    20d   4
    30d   4½
    40d   5

Construct a horizontal bar chart using the given data:

TITLE		DRAWING NO.
NAME	GRADE	

TITLE

DRAWING NO.

NAME

GRADE

Make a circle chart showing how your weekly allowance is spent.

Project Drawing and Print Reading

In pursuing your school studies, you may have an opportunity to take several shop courses that will involve activities in wood, metal, electricity, and other materials. All of these activities will require a certain amount of project planning. The process of planning any project consists of three major steps: (1) preparing a drawing, (2) determining the kind of materials to be used, and (3) listing the operations to be followed in constructing the project. The purpose of this unit is to give you some ideas in carrying out a project-planning assignment. It also describes several other types of project drawings.

Planning the project

Preparing a drawing. Once you have selected the particular project in which you plan to participate, you must prepare a drawing. Here you will have to decide on the kind of drawing that will best represent the construction details of the project. The drawing may be either a freehand sketch or a finished instrument drawing. For some relatively simple projects, a pictorial drawing may suffice. In most cases you will want to prepare a regular working drawing. When preparing such a drawing, use the basic drawing practices which you have learned in the previous units.

Preparing a bill of materials. After the drawing is completed, the next step is to identify the kinds and amounts of materials that will be needed for the project. Fig. 1 illustrates one type of form which is often used to list materials. Notice that

STUDENT PLANNING SHEET

Section I

Name of Project ______

Student Name ______ Date Started ______
Grade Level ______ Date Completed ______
Course ______ Instructor's Grade ______
Instructor's O.K. ______

Section II

Tools and Machines I will use

1. ______ 6. ______
2. ______ 7. ______
3. ______ 8. ______
4. ______ 9. ______
5. ______ 10. ______

Section III

Material I Will Need

No. Pieces	Description of Material	Size	Cost
		Total Cost	

Section IV

How I Plan To Make My Project

1. ______
2. ______
3. ______
4. ______
5. ______
6. ______
7. ______
8. ______
9. ______
10. ______
11. ______
12. ______
13. ______
14. ______

Fig. 1. A form used in project planning will usually contain a list of materials and a step-by-step procedure.

in this form you must specify the number of pieces required, type of materials, sizes, quantities, and cost.

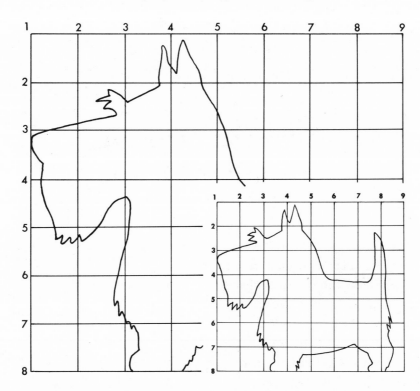

Fig. 2. Duplicating or enlarging a design may be done by using the grid method.

Duplicating a design

Preparing a work procedure. The final step in any project planning is to write out the work procedure which is to be followed in building the project. These operations are usually stated very briefly. The form shown in Fig. 1 is frequently used for planning the work procedure.

Duplicating a design

On some occasions you may want to duplicate a particular design, such as the silhouette of the dog in Fig. 2. Usually, in duplicating a design, the size will have to be increased. First draw cross-section lines over the illustration. Make the squares a definite size. Next prepare a cross section sheet with squares that will permit you to produce the object in the desired size.

For example, if the object is to be twice the size of the original, you may want to start with 1/4″ squares over the original illustration. Then in laying out the drawing sheet, you will need to use 1/2″ squares. Regular printed squared paper can be substituted as long as the given squares are used in their correct proportion.

Start drawing the figure on the large, squared sheet by duplicating the same line positions as on the original. See Fig. 2.

Sheet metal drawings

Many projects are made of sheet metal. Usually the shape of the sheet metal project is obtained by preparing a paper, cardboard, or metal pattern and then tracing it on the material to be used.

A pattern represents the actual flat layout of the object, that is, the unfolded shape of all of its surfaces. It is made by using one of several development techniques, depending on the shape of the object. For some projects, parallel line development is necessary, while others may require either radial line development or triangulation. For our purposes, we will describe only two methods: parallel line and radial line development.

Parallel line development. Parallel line development is used when the surfaces of the object are parallel, such as in a square or rectangular box or pipe.

Two views are required to draw a development

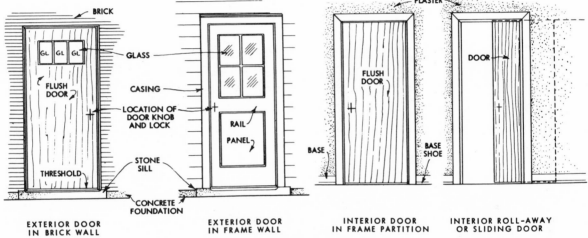

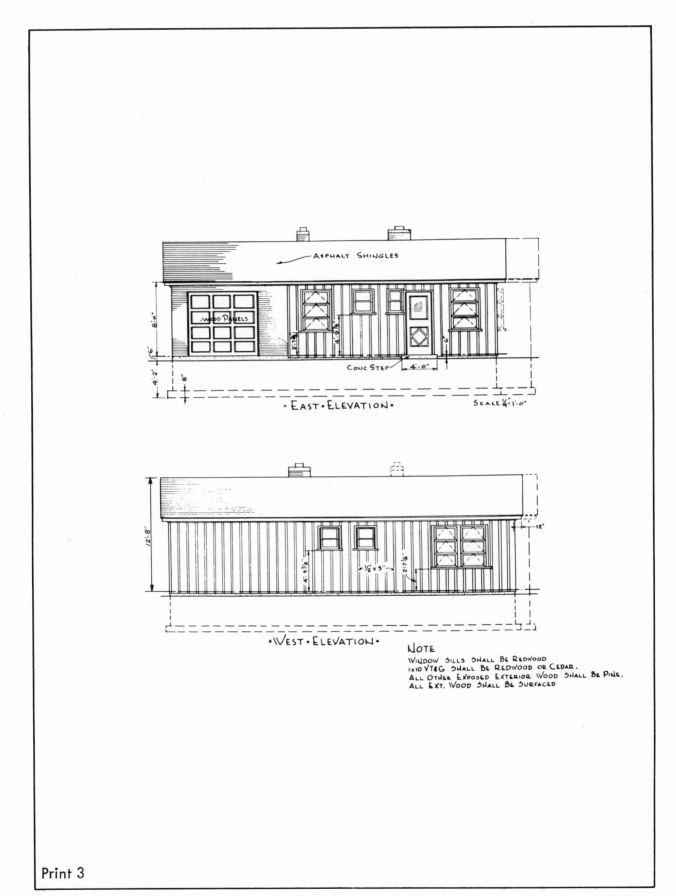

· EAST · ELEVATION ·   Scale ¼"=1'-0"

· WEST · ELEVATION ·

NOTE

WINDOW SILLS SHALL BE REDWOOD
1x10 VT&G SHALL BE REDWOOD OR CEDAR.
ALL OTHER EXPOSED EXTERIOR WOOD SHALL BE PINE.
ALL EXT. WOOD SHALL BE SURFACED

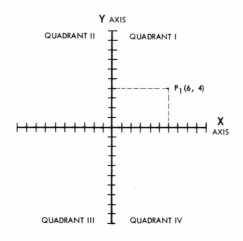

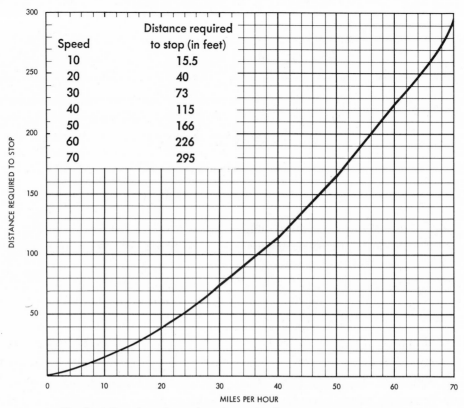

Speed	Distance required to stop (in feet)
10	15.5
20	40
30	73
40	115
50	166
60	226
70	295

Fig. 2. A line chart shows the relationship between two sets of data.

plotted in relation to the speed of the car. The two sets of data are the distance, in feet, required to stop the car, and the speed of the car. The speed of the car represents the more regular, or easily controlled, of the two sets of data. The speed will be plotted along the horizontal axis. The stopping

distance is the more variable of the two sets of data, and will be plotted along the vertical axis.

Examining the two sets of data, it is seen that the speed ranges from 10 mph to 70 mph. Therefore the horizontal axis will be scaled from 0 to 70 mph, each unit along the X-axis representing

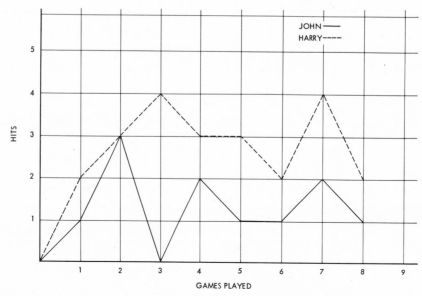

Fig. 3. Several sets of figures can be compared on one chart.

Hour	Temperature
7 A.M.	60°
8 A.M.	65°
9 A.M.	68°
10 A.M.	70°
11 A.M.	74°
12 P.M.	80°
1 P.M.	84°
2 P.M.	86°
3 P.M.	90°
4 P.M.	70°
5 P.M.	68°
6 P.M.	60°
7 P.M.	55°

Construct a line chart showing the above hourly temperatures. Use any desired units.

TITLE

NAME

GRADE

DRAWING NO.

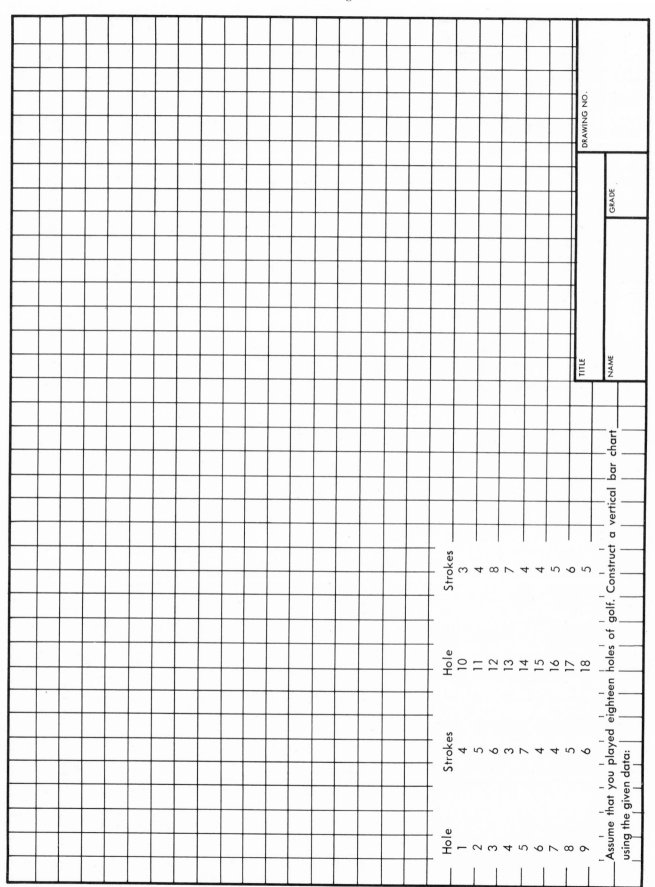

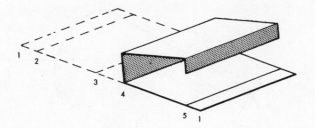

PATTERN BEING FORMED INTO SHAPE

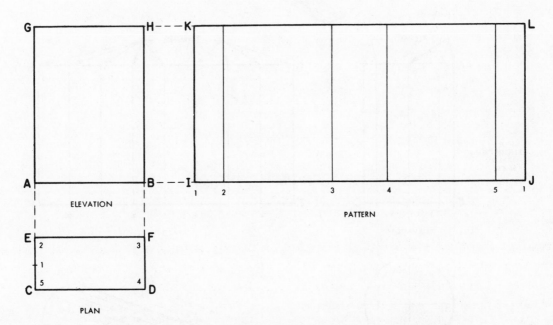

ELEVATION

PLAN

PATTERN

Fig. 3. The pattern is developed from the elevation and plan views.

for the rectangular pipe in Fig. 3. One view is needed to show the height of the pipe; a second is needed to show the sides of the pipe. Any view front or side, which shows the height of a project is called an *elevation view*. In Fig. 3 sides AG and HB represent the height of the pipe and the height necessary for its development.

A cross-section of the pipe is also needed to show the sum of the sides, or circumference of the pipe. This cross-section is called a *plan view*. The pattern or development will be folded to the shape shown in the cross-section of the pipe. The length of the development is determined by the sum of the sides seen in the cross section. Thus, lines KL and IJ in the development will equal the sum of sides EC, CD, DF, and FE in the plan view. The necessary dimensions for the development have now been established.

The seam is the point at which the ends of the pattern will meet when the pattern is folded as shown in the plan view. In the plan view of Fig. 3, the seam is numbered as point 1, and the corners of the pipe are numbered as shown.

Having determined the height and length of the pattern it is now possible to draw the pattern. Lines KL and IJ are extensions of the top and bottom of the pipe seen in the elevation. As already shown, the length of KL or IJ is equal to the sum of the sides shown in the cross-section, or plan view. KL and IJ are called stretchout lines. The numbered points of the stretchout line correspond to the numbered points of the plan view. Thus, lengths 1-2, 2-3, 3-4, 4-5, and 5-1 of the stretchout line correspond to lengths of 1-2, 2-3, etc. of the plan view.

The lines drawn from the numbered points are

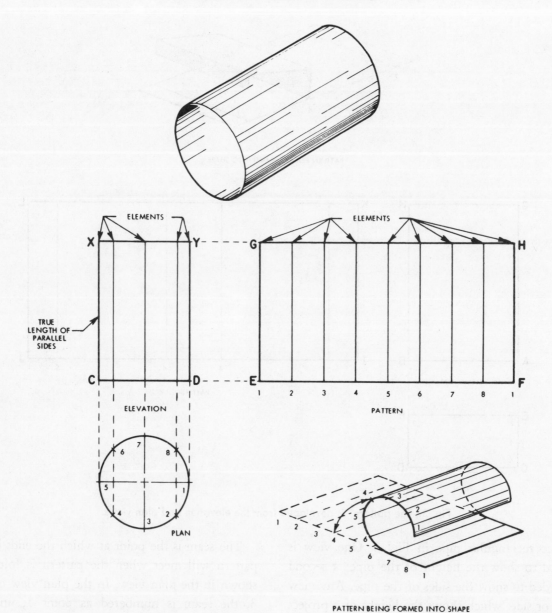

Fig. 4. The length of the stretchout line is determined by the circumference of the object seen in the plan view.

called *elements*. They are perpendicular to the stretchout lines and indicate in this case where the development is to be folded. The elements complete the development and represent the corners of the pipe seen in the plan view.

Fig. 4 shows the method for the development of a circular pipe, or cylinder. The plan view is drawn first, and appears as a circle. The circle is divided into an even number of equal parts. In Fig. 4, the circle is divided into eight parts.

The elevation view is drawn directly above the plan view. The true height of the cylinder, lines XC and YD, is shown. Extending lines XY and CD to the right will give the true height of the development.

The extension of line CD to the development, line EF, will be the stretchout line. Starting at point 1 on the plan view, lengths 1-2, 2-3, 3-4, etc., are transferred to the stretchout line EF beginning at point E until the last length, 8-1, has been transferred. The full length of the stretchout line is thus determined. The development is fin-

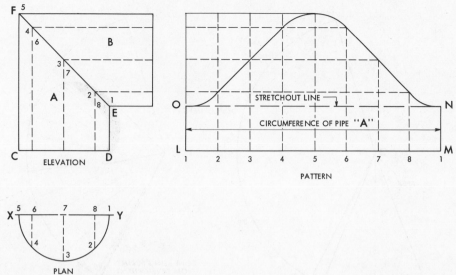

Fig. 5. The irregular curve of the pattern is found by projecting points from the elevation view to elements of the pattern.

ished by drawing the elements at the required points on the stretchout line.

Fig. 5 illustrates the need for elements in drawing a development. In many cases, the object to be drawn is shown cut at an angle or joining other objects, as in Fig. 5. The development when drawn will not appear as a square or rectangle; it will appear as an irregularly curved surface.

The plan view is again drawn first. Note that in Fig. 4, equally spaced points in the lower half-circle of the plan view coincide with the points in the upper half when the points are projected up from the plan to the elevation view. Therefore, instead of using a full plan (circle) we need only use a half-plan (half-circle) to determine where all the points will intersect the elevation view. To determine the position of these points, proceed as follows:

Divide the half-plan into an even number of equal lengths. Number the points as shown, starting with point 1 on line XY and concluding with points 6, 7 and 8 on the horizontal axis line. Points 2, 3, and 4 will coincide with points 8, 7 and 6 when projected to the elevation view.

Directly above the plan view, construct the elevation view. This view will show the angle of the cut, or the angle at which objects such as pipe A and pipe B are to be joined. In this problem, however, we are concerned only with the development

of pipe A. Project the numbered points of the plan view up to the line showing the intersection of pipes A and B (line FE) in the elevation view.

Extend line CD and point E to the right. The line LM will be the bottom of the development. For this problem, the extension line from point E (line ON) will be the stretchout line. Transfer the equally spaced lengths on the half-circle in the plan view to the stretchout line ON in the development. From the eight points on the stretchout line, construct lines perpendicular to the stretchout line. These lines are the elements.

From the points where the projections from the plan view meet the line of intersection FE, draw parallel horizontal lines to the development. Thus points 2 and 8 are projected to FE. They meet FE at the same point and are then projected to the development. Where the projection of points 2 and 8 from line FE intersects elements from points 2 and 8 on the stretchout line, mark the points of intersection. The same procedure will be used with points 1, 3-7, 4-6, and 5, until 8 points have been determined for the curve that represents the unknown shape of the development. Connect the points with an irregular curve to complete the development.

Radial line development. Another way of drawing the pattern, or development, for a sheet metal object is to use the radial line development

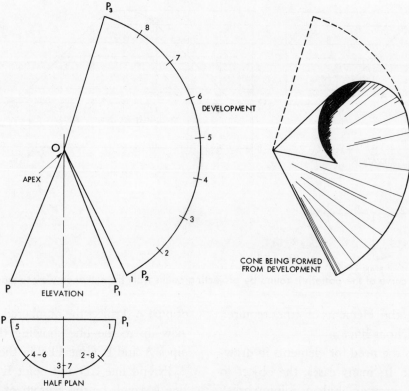

Fig. 6. Drawing the pattern will be simplified if the elevation and pattern views use the same point of origin.

method. This method is used only when the object to be constructed has regular tapered sides, such as a pyramid, a cone or part of a cone.

A cone is a figure which has a circular base and a surface which slopes inward or tapers to a point called the apex. The method for developing a cone-like object is called "radial" because a conical figure can be imagined to be formed of many lines "radiating" out from the apex to the circular base. The conical shape finds many uses; from pop corn containers to nose cones, and from ice cream cones to funnels.

The radial line development method also uses the plan, or half-plan, and elevation views to draw the development. In Fig. 6, the half-plan and elevation views are first established. Lines OP and OP_1 in the elevation view give the true length of the "sides" of the cone.

From our experience with parallel line development, we know that the length of the "side" of the cone seen in the elevation view is transferred to the development. The lengths of the "sides" are always equal. The "side" of the development, line OP_2 is

drawn, equal to lines OP and OP_1. For convenience, line OP_2 of the development is drawn alongside line OP_1 of the elevation view.

With OP_2 as a radius, lightly draw an arc to represent the stretchout line. The stretchout line will be circular because the object is a cone and any point on the base of the cone—the stretchout line is the base—will be an equal distance from the origin point O. Notice that the elevation and the development views use the same point of origin, point O.

We have found the length of the side of the development and the shape and direction of the stretchout line. To complete the development, we must now find the length of the stretchout line.

As in the parallel line development method, the half-circle of the half-plan is divided into an even number of equal lengths. The lengths are then transferred to the stretchout line. The total length of the stretchout line is equal to the sum of the lengths of the circle, or to the circumference of the plan view circle. After the length of the stretchout line has been established, complete the

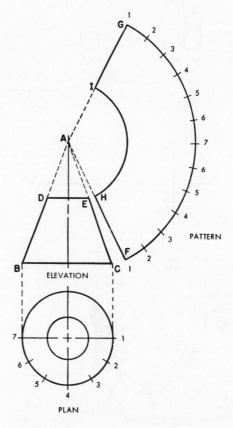

Fig. 7. Procedure for constructing the pattern view of any cone or part of a cone is essentially identical.

drawing of the development by drawing line OP₃.

Fig. 7 illustrates the method of drawing a development for a section of a cone. Such a section might be used for a funnel. Notice that the procedure is identical to the procedure for the development of a regular cone.

In Fig. 7, line AF of the development is drawn alongside and equal to line AC of the elevation view. Similarly, line AH is made equal to line AE. The stretchout line is drawn in the development with AF as a radius and point A as origin. The top of the section of the cone is drawn in the development view with AH as a radius and point A as origin.

Notice that it is not necessary to find the length of arc HI. The arc is bounded by the lines AF and AG. Line AG is drawn when the length of the stretchout line is found. When line AG is drawn, it represents the end of arc HI as well as the end of the stretchout line or arc FG.

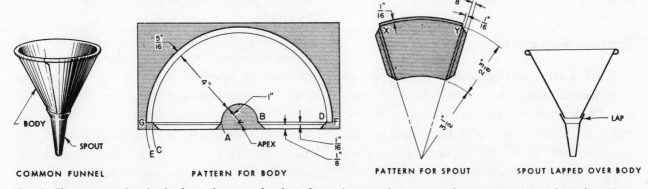

Fig. 8. The patterns for the body and spout of a funnel are shown with provisions for an upper rim and overlap where body and spout meet.

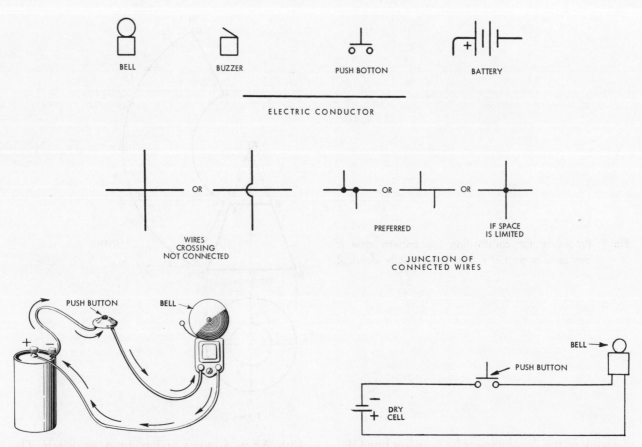

Fig. 9. Notice that the schematic for the door bell can be drawn more quickly and more easily than the actual picture of the door bell.

Fig. 8 shows the patterns for a funnel. Notice that additional metal is provided to allow for the folding of a rim at the top of the funnel. Overlap where the body and spout join is also provided for.

The sheet metal drawing is usually fully dimensioned, with dimensions given in inches. Give all information necessary for the completion of the project. Use notes to give information that cannot be included in dimensions.

Electrical diagrams

Today, a large part of industry is composed of firms manufacturing electrical and electronic products. Clocks, can openers, tooth brushes, electric knives, drills, saws, vacuum cleaners, floor polishers,—even the latest toys are operated by electricity. There is an excellent chance that you will someday find yourself in a position which requires at least a basic knowledge of simple electrical wiring and wiring diagrams. Wiring diagrams are usually referred to as *schematics*.

The tremendous growth of the electric and electronics industry has opened up many opportunities in the field of electricity. Schools recognize and try to meet the genuine need for persons trained in various electrical fields. Many schools offer courses which provide the student with at least a basic knowledge of wiring circuits and reading electrical blueprints.

A schematic is essentially a diagram which shows the various parts of an electric circuit and the flow of electricity through that circuit. The schematic uses lines and symbols to represent the actual wiring and the components of the circuit. Schematics are seldom, if ever, drawn to scale; the major concern is not with the actual physical installation of the circuit, but rather with the arrangement of the necessary components to do a specific job.

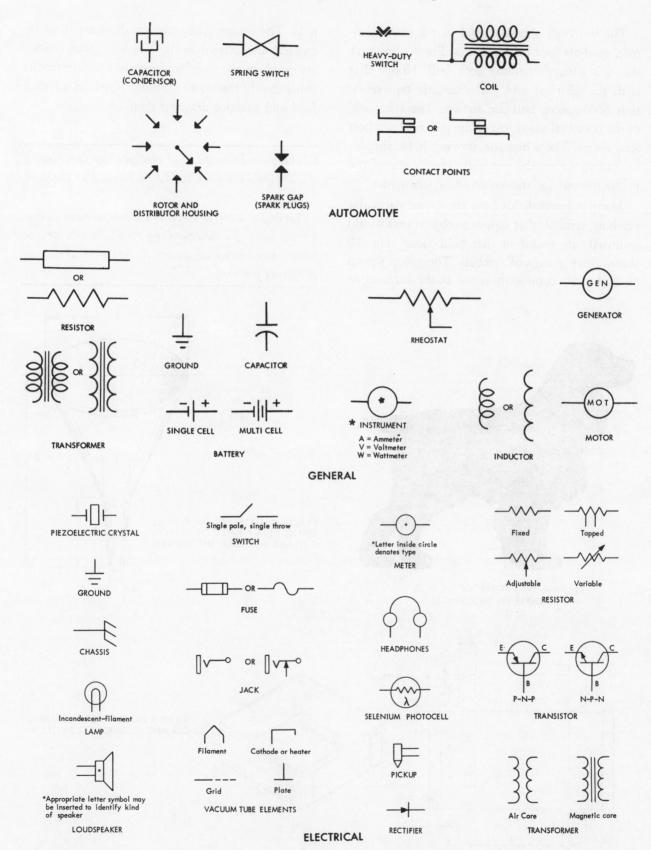

CAPACITOR (CONDENSOR)

SPRING SWITCH

HEAVY-DUTY SWITCH

COIL

ROTOR AND DISTRIBUTOR HOUSING

SPARK GAP (SPARK PLUGS)

CONTACT POINTS

AUTOMOTIVE

RESISTOR

TRANSFORMER

GROUND

CAPACITOR

SINGLE CELL

MULTI CELL

BATTERY

RHEOSTAT

GENERATOR

* **INSTRUMENT**
A = Ammeter
V = Voltmeter
W = Wattmeter

INDUCTOR

MOTOR

GENERAL

PIEZOELECTRIC CRYSTAL

Single pole, single throw
SWITCH

*Letter inside circle denotes type
METER

Fixed   Tapped

Adjustable   Variable
RESISTOR

GROUND

FUSE

HEADPHONES

CHASSIS

JACK

λ
SELENIUM PHOTOCELL

E   C   E   C

B   B
P-N-P   N-P-N
TRANSISTOR

Incandescent-filament
LAMP

Filament   Cathode or heater

Grid   Plate
VACUUM TUBE ELEMENTS

PICKUP

Air Core   Magnetic core
TRANSFORMER

*Appropriate letter symbol may be inserted to identify kind of speaker
LOUDSPEAKER

RECTIFIER

ELECTRICAL

Fig. 10. Some schematic symbols are used more in one field than in another. Most symbols are general and used by all fields.

Fig. 9 (Top) shows several of the more common symbols used in schematics. Fig 9 (Bottom) shows a battery operated door bell. Notice that both the pictorial and the schematic representation of the door bell are shown. The schematic of the door bell uses most of the electrical symbols seen above. The schematic drawing is far simpler to construct than the pictorial. This is one of the prime reasons for the existence of schematics.

Many different fields have their own particular symbols, symbols that represent objects that would ordinarily be found in that field alone. Fig. 10 shows three groups of symbols. The group shown at the top are commonly found in the automotive field. The center group are usually used in architectural and related fields. The last group shown are used in the communications and electronics fields. Study the more common symbols of each field and practice drawing them.

Check your knowledge of this unit by completing the Shelf-Check for Unit 12 on pages 165 through 171.

Several pattern projects are recommended below which may be assigned by your instructor, or which you may wish to try to test your ability to construct patterns.

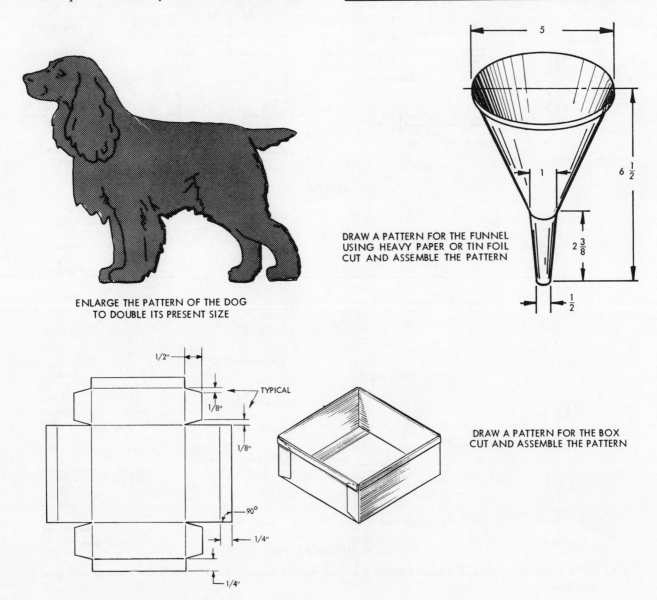

ENLARGE THE PATTERN OF THE DOG
TO DOUBLE ITS PRESENT SIZE

DRAW A PATTERN FOR THE FUNNEL
USING HEAVY PAPER OR TIN FOIL
CUT AND ASSEMBLE THE PATTERN

DRAW A PATTERN FOR THE BOX
CUT AND ASSEMBLE THE PATTERN

DIRECTIONS: Identify the symbols shown below. Write the name for each in the space provided.

1. _____   2. _____   3. _____

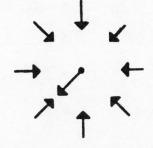

4. _____   5. _____   6. _____

7. _____   8. _____   9. _____

10. _____   11. _____   12. _____

Score_____

PART 2

DIRECTIONS: Using Drawing 1, p. 167, answer the questions listed below.

Questions   Answers

1. How many different components are shown in the schematic?   1. _____

2. Is the antenna coil grounded?   2. _____

3. How many resistors are shown?   3. _____

4. How many capacitors?   4. _____

5. What is the power source?   5. _____

6. What is the voltage of the power source?   6. _____

7. How many wires are connected to the transistor?   7. _____

DIRECTIONS: Using Drawing 2, p. 167, answer the questions listed below.

Questions   Answers

8. How many switches are shown?   1. _____

9. How many coils are shown?   2. _____

10. What components are grounded?   3. _____

11. Will closing the ignition switch start the starter motor?   4. _____

12. What side of the battery is grounded?   5. _____

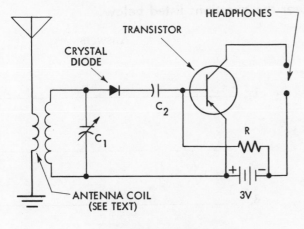

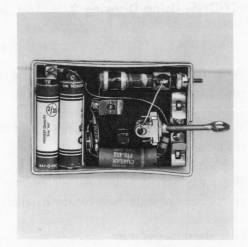

SCHEMATIC DIAGRAM

DRAWING 1

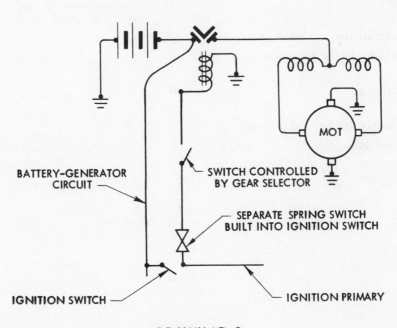

DRAWING 2

PART 3

DIRECTIONS: Using Drawing 3, p. 169, answer the questions listed below.

Questions   Answers

1. What type of steel is used to make this cross
   peen hammer?   1. _____

2. What are the dimensions of the piece of stock
   required to make this hammer?   2. _____

3. What size drill is to be used for making the
   eye? (The eye is the hole for the handle.)   3. _____

4. Give the dimension B.   4. _____

5. What is the radius of the rounded end of the
   peen?   5. _____

6. Give the dimension C.   6. _____

7. What type of line represents the eye in the
   right-side view?   7. _____

8. Surface F, shown in the top view, is represented by what letter in the front view?   8. _____

9. Give the dimension E.   9. _____

10. What is the distance between the centers of
    the 5/16" drilled holes?   10. _____

Score_____

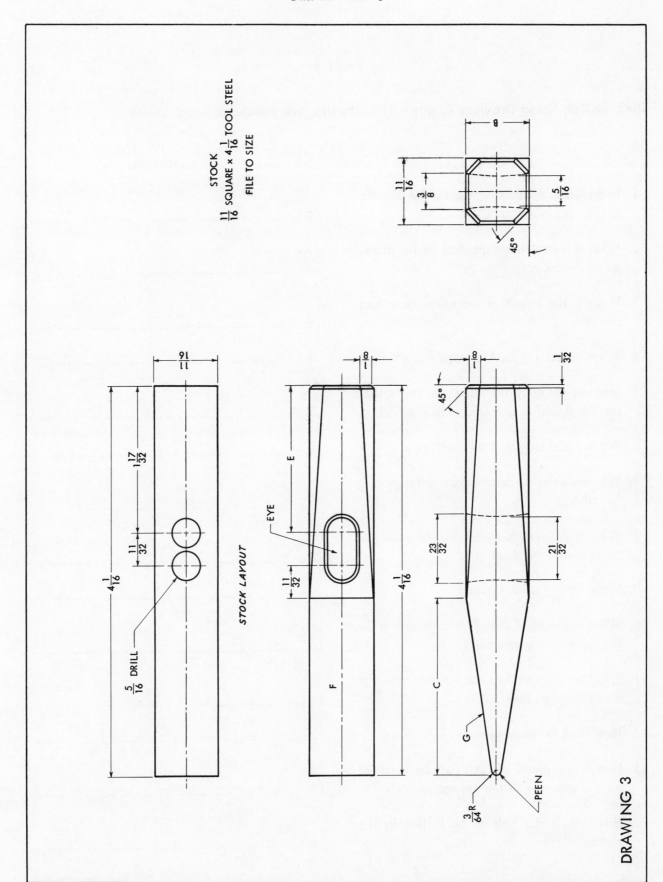

STOCK
$\frac{11}{16}$ SQUARE × $4\frac{1}{16}$ TOOL STEEL

FILE TO SIZE

$\frac{11}{16}$

$\frac{3}{8}$

$\frac{5}{16}$

45°

STOCK LAYOUT

$\frac{11}{16}$

$1\frac{17}{32}$

$\frac{11}{32}$

$4\frac{1}{16}$

$\frac{5}{16}$ DRILL

EYE

E

$\frac{1}{8}$

$\frac{11}{32}$

$4\frac{1}{16}$

F

$\frac{1}{8}$

$\frac{1}{32}$

45°

$\frac{23}{32}$

$\frac{21}{32}$

C

G

$\frac{3}{64}$R

PEEN

DRAWING 3

PART 4

DIRECTIONS: Using Drawing 4, page 171, answer the questions listed below.

Questions   Answers

1. In the assembly drawing, what is the diameter
of the body of the tool?   1. _____

2. What is the tolerance specified for the diameter of the body?   2. _____

3. What is the length of the roller, part number 4?   3. _____

4. What material is used for parts 4 and 6?   4. _____

5. How are the body, the roller, and the washer,
parts 1, 4, and 6, indicated to be finished?   5. _____

6. What is the length of the body, part 1?   6. _____

7. Why are there no detail drawings for parts 2,
3, and 5?   7. _____

8. What clearance is called for between the
roller and the washer?   8. _____

9. What type of screw is part 5?   9. _____

10. What is the size of the tapped hole seen in the
side view of the assembly?   10. _____

11. What are the specifications for the screw that
fits this tapped hole?   11. _____

12. How thick is the washer?   12. _____

13. How many pieces of each part are required
for the assembly of the trepanning tools?   13. _____

14. How big is the hole to be drilled in the
washer?   14. _____

Score_____

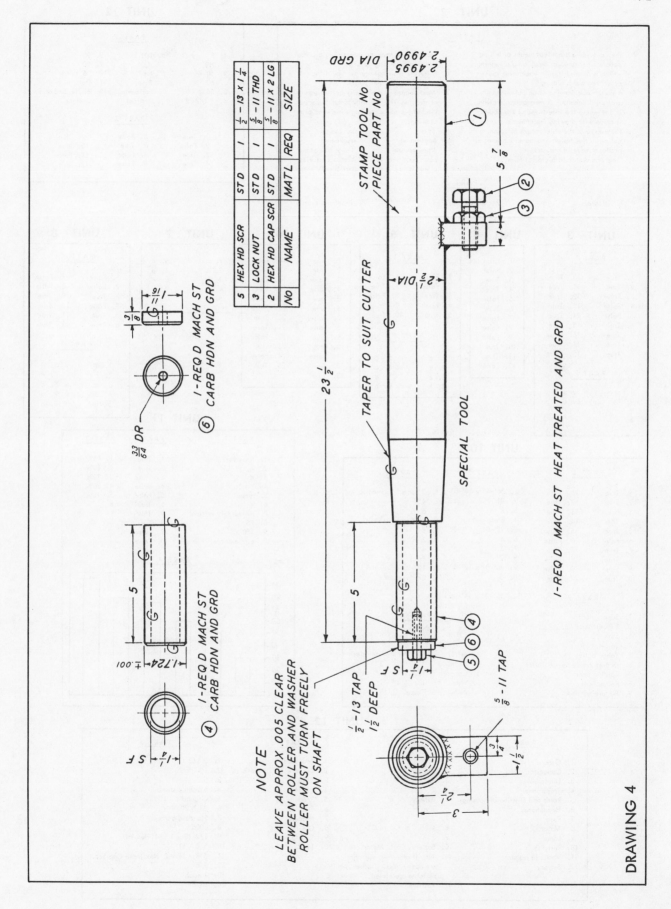

No	NAME	MATL	REQ	SIZE
5	HEX HD SCR	ST D	1	$\frac{1}{2}$ – 13 x 1$\frac{1}{4}$
3	LOCK NUT	ST D	1	$\frac{5}{8}$ – 11 THD
2	HEX HD CAP SCR	ST D	1	$\frac{5}{8}$ – 11 x 2 LG

STAMP TOOL No
PIECE PART No

2.4995 / 2.4990 DIA GRD

5$\frac{7}{8}$

2$\frac{1}{2}$ DIA

23$\frac{1}{2}$

TAPER TO SUIT CUTTER

SPECIAL TOOL

1 – REQ D MACH ST HEAT TREATED AND GRD

5

$\frac{1}{2}$ – 13 TAP
1$\frac{1}{2}$ DEEP

1$\frac{1}{4}$ S F

$\frac{5}{8}$ – 11 TAP

③ ④ ⑤ ⑥

⑥ $\frac{33}{64}$ DR

$\frac{3}{8}$
1$\frac{11}{16}$

1 – REQ D MACH ST
CARB HDN AND GRD

④ 1$\frac{1}{4}$ S F

1.724 ±.001

5

1 – REQ D MACH ST
CARB HDN AND GRD

NOTE
LEAVE APPROX .005 CLEAR
BETWEEN ROLLER AND WASHER
ROLLER MUST TURN FREELY
ON SHAFT

$\frac{3}{4}$

1$\frac{1}{2}$

2$\frac{1}{4}$

3

DRAWING 4

Answer Key to Self-Check-Questions

UNIT 1

PART 1

1. They provide a graphical means of telling people what to do or describing something which is often difficult when using spoken or written words alone.
2. A sketch is a drawing of an object made entirely freehand. A mechanical drawing is one where the true shape of the object is made with drawing instruments.
3. They serve as the chief means of communication in the design and manufacture of all products made in industry.
4. Freehand sketch in pictorial form.
5. Mechanical drawing containing all essential manufacturing details.
6. Different drawings are used by different people for different purposes.
7. Most people in their daily activities are confronted with the task of making some kind of sketch or reading some kind of a drawing.
8. Toys
   Appliances
   Wheelbarrow
   Lawn furniture
   Light fixture
   Bicycle
   Workbench
   X-mas tree stand
   Auto accessories
   Portable TV stand

PART 2

1. Diagram
2. Mechanical drawing
3. Freehand sketch

UNIT 2

PART 1

1. F	5. left	8. 90
HB	right	9. 90
2. 50	6. top	10. 6
60	down	11. flat
3. 30	7. horizontal	12. circles and arcs
4. 1 1/2 - 2	vertical	straight

PART 2

1. True	5. False	9. True
2. False	6. True	10. True
3. True	7. False	11. False
4. True	8. False	12. True

UNIT 3

PART 1

1. True	10. True
2. True	11. True
3. True	12. True
4. False	13. True
5. False	14. True
6. True	15. True
7. False	16. False
8. True	17. True
9. True	18. False

PART 2

1. c	5. a
2. b	6. b
3. c	7. c
4. d	

UNIT 4

1. False
2. False
3. True
4. True
5. False
6. True
7. False
8. False
9. False
10. True
11. True
12. False

UNIT 5

PART 1

1. False	6. False
2. False	7. True
3. True	8. False
4. False	9. True
5. True	10. True

PART 2

1. c	5. c
2. b	6. c
3. c	7. d
4. b	8. a

UNIT 6

1. True	14. False
2. True	15. True
3. False	16. True
4. False	17. True
5. True	18. False
6. True	19. False
7. False	20. True
8. False	21. True
9. False	22. True
10. True	23. True
11. False	24. False
12. True	25. True
13. True	

UNIT 7

1. b	6. b
2. c	7. a
3. b	8. c
4. d	9. c
5. a	10. b

UNIT 8

PART 1

1. False	7. False
2. False	8. False
3. True	9. True
4. False	10. True
5. True	11. True
6. True	12. True

PART 2

1. d	6. d
2. d	7. d
3. c	8. c
4. d	9. d
5. a	10. a

UNIT 10

PART 1

1. True	11. True
2. True	12. True
3. False	13. True
4. True	14. True
5. True	15. True
6. False	16. False
7. True	17. True
8. True	18. False
9. False	19. False
10. True	20. True

PART 2

1. Supply duct
2. Exhaust duct
3. Sink
4. Toilet
5. Range outlet
6. Duplex convenience outlet
7. Folding door
8. Frame wall
9. Picture window

PART 3

1. 36'-4" x 64'-0"
2. 21
3. 8
4. Stairs--down
5. 12'-0" x 17'-0"
6. 4
7. 3
8. 2
9. 10'-0" x 12'-4"
10. 5
11. 1
12. 12'-0" x 14'-4"

PART 4

1. 2 doors, 6 windows
2. 4"
3. 6'-8"
4. Sliding glass door
5. 8
6. 8'-0"
7. 4'-8"
8. Brick
9. 12"
10. Concrete

PART 5

1. 12'-8"
2. 4'-9 7/8"
3. Brick
4. Redwood
5. Concrete
6. Wood
7. Redwood or Cedar
8. Pine
9. 6"
10. Asphalt shingles
11. 1/2" x 3"
12. 2'-7 1/4"
13. 1/4" = 1'

UNIT 11

PART 1	PART 2	PART 3
1. line, bar	1. True	1. 15%
2. X, Y	2. True	2. 10%
3. X	3. False	3. 22%
4. Y	4. False	4. 30%
5. range	5. False	5. 18%
6. Y	6. True	6. 5%
7. horizontal, vertical	7. True	
8. 25	8. True	
9. 180	9. False	
10. circle or pie	10. False	

PART 4

UNIT 12

PART 1

1. Ground
2. Capacitor
3. Loudspeaker
4. Transformer
5. Rotor and Distributor housing
6. Headphones
7. Fuse
8. Motor
9. Lamp
10. Resistor
11. Variable resistor
12. Battery

PART 2

1. 11
2. Yes
3. 1
4. 2
5. Battery
6. 3 volts
7. 3
8. 4
9. 3
10. Battery, Coil, Motor
11. No, if other two switches are open. Yes, if other two switches are closed.
12. Positive

PART 3

1. Tool steel
2. 11/16 x 11/16 x 4 1/16
3. 5/16
4. 11/16
5. 3/64
6. 1 27/32
7. Hidden
8. G
9. 1 17/32
10. 11/32

PART 4

1. 2 1/2
2. .0005"
3. 5"
4. Machine steel
5. Ground
6. 23 1/2"
7. Standard parts
8. .005"
9. 1/2-13 x 1 1/4 Hex HD
10. 5/8 - 11
11. 5/8 - 11 x 2 Hex Head Cap Scr.
12. 3/8
13. 1 each
14. 33/64